The Trouble with Physics

The
TROUBLE
with PHYSICS

The Rise of String Theory,
the Fall of a Science,
and What Comes Next

Lee Smolin

HOUGHTON MIFFLIN COMPANY
BOSTON • NEW YORK

To Kai

For information about permission to reproduce selections from
this book, write to Permissions, Houghton Mifflin Company,
215 Park Avenue South, New York, New York 10003.

Visit our Web site: www.houghtonmifflinbooks.com.

Library of Congress Cataloging-in-Publication Data
Smolin, Lee, date.
The trouble with physics : the rise of string theory,
the fall of a science, and what comes next / Lee Smolin.
p. cm.
Includes bibliographical references and index.
ISBN-13: 978-0-618-55105-7
ISBN-10: 0-618-55105-0
1. Physics — Methodology — History — 20th century.
2. String models. I. Title.
QC6.S6535 2006
530.14—dc22 2006007235

Printed in the United States of America

Book design by Robert Overholtzer
Illustrations by Michael Prendergast

MP 10 9 8 7 6 5 4 3

Contents

Introduction

There may or may not be a God. Or gods. Yet there is something ennobling about our search for the divine. And also something humanizing, which is reflected in each of the paths people have discovered to take us to deeper levels of truth. Some seek transcendence in meditation or prayer; others seek it in service to their fellow human beings; still others, the ones lucky enough to have the talent, seek transcendence in the practice of an art.

Another way of engaging life's deepest questions is science. Not that every scientist is a seeker; most are not. But within every scientific discipline, there are those driven by a passion to know what is most essentially true about their subject. If they are mathematicians, they want to know what numbers are, or what kind of truth mathematics describes. If they are biologists, they want to know what life is, and how it started. If they are physicists, they want to know about space and time, and what brought the world into existence. These fundamental questions are the hardest to answer and progress is seldom direct. Only a handful of scientists have the patience for this work. It is the riskiest kind of work, but the most rewarding: When someone answers a question about the foundations of a subject, it can change everything we know.

Because it is their job to add to our growing store of knowledge, scientists spend their days confronting what they don't understand.

And those scientists who work on the foundations of any given field are fully aware that the building blocks are never as solid as their colleagues tend to believe.

This is the story of a quest to understand nature at its deepest level. Its protagonists are the scientists who are laboring to extend our knowledge of the basic laws of physics. The period of time I will address — roughly since 1975 — is the span of my own professional career as a theoretical physicist. It may also be the strangest and most frustrating period in the history of physics since Kepler and Galileo began the practice of our craft four hundred years ago.

The story I will tell could be read by some as a tragedy. To put it bluntly — and to give away the punch line — we have failed. We inherited a science, physics, that had been progressing so fast for so long that it was often taken as the model for how other kinds of science should be done. For more than two centuries, until the present period, our understanding of the laws of nature expanded rapidly. But today, despite our best efforts, what we know for certain about these laws is no more than what we knew back in the 1970s.

How unusual is it for three decades to pass without major progress in fundamental physics? Even if we look back more than two hundred years, to a time when science was the concern mostly of wealthy amateurs, it is unprecedented. Since at least the late eighteenth century, significant progress has been made on crucial questions every quarter century.

By 1780, when Antoine Lavoisier's quantitative chemistry experiments were showing that matter is conserved, Isaac Newton's laws of motion and gravity had been in place for almost a hundred years. But while Newton gave us a framework for understanding all of nature, the frontier was wide open. People were just beginning to learn the basic facts about matter, light, and heat, and mysterious phenomena like electricity and magnetism were being elucidated.

Over the next twenty-five years, major discoveries were made in each of these areas. We began to understand that light is a wave. We discovered the law that governs the force between electrically charged particles. And we made huge leaps in our understanding of matter with John Dalton's atomic theory. The notion of energy was introduced; interference and diffraction were explained in terms of the wave theory of light; electrical resistance and the relationship between electricity and magnetism were explored.

Several basic concepts underlying modern physics emerged in the next quarter century, from 1830 to 1855. Michael Faraday introduced the notion that forces are conveyed by fields, an idea he used to greatly advance our understanding of electricity and magnetism. During the same period, the conservation of energy was proposed, as was the second law of thermodynamics.

In the quarter century following that, Faraday's pioneering ideas about fields were developed by James Clerk Maxwell into our modern theory of electromagnetism. Maxwell not only unified electricity and magnetism, he explained light as an electromagnetic wave. In 1867, he explained the behavior of gases in terms of the atomic theory. During the same period, Rudolf Clausius introduced the notion of entropy.

The period from 1880 to 1905 saw the discoveries of electrons and X rays. The study of heat radiation was developed in several steps, leading to Max Planck's discovery, in 1900, of the right formula to describe the thermal properties of radiation — a formula that would spark the quantum revolution.

In 1905, Albert Einstein was twenty-six. He had failed to find an academic job in spite of the fact that his early work on the physics of heat radiation alone would come to be seen as a major contribution to science. But that was just a warm-up. He soon zeroed in on the fundamental questions of physics: First, how could the relativity of motion be reconciled with Maxwell's laws of electricity and magnetism? He told us in his special theory of relativity. Should we think of the chemical elements as Newtonian atoms? Einstein proved we must. How can we reconcile the theories of light with the existence of atoms? Einstein told us how, and in the process showed that light is both a wave and a particle. All in the year 1905, in time stolen from his work as a patent examiner.

The working out of Einstein's insights took the next quarter century. By 1930, we had his general theory of relativity, which makes the revolutionary claim that the geometry of space is not fixed but evolves in time. The wave-particle duality uncovered by Einstein in 1905 had become a fully realized quantum theory, which gave us a detailed understanding of atoms, chemistry, matter, and radiation. By 1930 we also knew that the universe contained huge numbers of galaxies like our own, and we knew they were moving away from one another. The implications were not yet clear, but we knew we lived in an expanding universe.

With the establishment of quantum theory and general relativity as part of our understanding of the world, the first stage in the twentieth-century revolution in physics was over. Many physics professors, uncomfortable with revolutions in their areas of expertise, were relieved that we could go back to doing science the normal way, without having to question our basic assumptions at every turn. But their relief was premature.

Einstein died at the end of the next quarter century, in 1955. By then we had learned how to consistently combine quantum theory with the special theory of relativity; this was the great accomplishment of the generation of Freeman Dyson and Richard Feynman. We had discovered the neutron and the neutrino and hundreds of other apparently elementary particles. We had also understood that the myriad phenomena in nature are governed by just four forces: electromagnetism, gravity, the strong nuclear force (which holds atomic nuclei together), and the weak nuclear force (responsible for radioactive decay).

Another quarter century brings us to 1980. By then we had constructed a theory explaining the results of all our experiments on the elementary particles and forces to date — a theory called the standard model of elementary-particle physics. For example, the standard model told us precisely how protons and neutrons are made up of quarks, which are held together by gluons, the carriers of the strong nuclear force. For the first time in the history of fundamental physics, theory had caught up with experiment. No one has since done an experiment that was not consistent with this model or with general relativity.

Going from the very small to the very large, our knowledge of physics now extended to the new science of cosmology, where the Big Bang theory had become the consensus view. We realized that our universe contains not only stars and galaxies but exotic objects such as neutron stars, quasars, supernovas, and black holes. By 1980, Stephen Hawking had already made the fantastic prediction that black holes radiate. Astronomers also had evidence that the universe contains a lot of dark matter — that is, matter in a form that neither emits nor reflects light.

In 1981, the cosmologist Alan Guth proposed a scenario for the very early history of the universe called inflation. Roughly speaking, his theory asserts that the universe went through a spurt of enormous growth extremely early in its life, and it explains why the universe looks pretty much the same in every direction. The theory of inflation made predictions that seemed dubious, until the evidence began to swing toward them a decade ago. As of this writing, a few puzzles remain, but the bulk of the evidence supports the predictions of inflation.

Thus, by 1981, physics had enjoyed two hundred years of explosive growth. Discovery after discovery deepened our understanding of nature, because in each case theory and experiment had marched hand in hand. New ideas were tested and confirmed and new experimental discoveries were explained in terms of theory. Then, in the early 1980s, things ground to a halt.

I am a member of the first generation of physicists educated since the standard model of particle physics was established. When I meet old friends from college and graduate school, we sometimes ask each other, "What have we discovered that our generation can be proud of?" If we mean new fundamental discoveries, established by experiment and explained by theory — discoveries on the scale of those just mentioned — the answer, we have to admit, is *"Nothing!"* Mark Wise is a leading theorist working on particle physics beyond the standard model. At a recent seminar at the Perimeter Institute of Theoretical Physics, in Waterloo, Ontario, where I work, he talked about the problem of where the masses of the elementary particles come from. "We've been remarkably unsuccessful at solving that problem," he said. "If I had to give a talk on the fermion-mass problem now, I'd probably end up talking about things I could have in the 1980s."[1] He went on to tell a story about when he and John Preskill, another leading theorist, arrived at Caltech in 1983, to join its faculty. "John Preskill and I were sitting together in his office, talking. . . . You know, the gods of physics were at Caltech, and now we were there! John said, 'I'm not going to forget what is important to work on.' So he took what was known about the quark and lepton masses, and he wrote it on a yellow sheet of paper and stuck it on his bulletin board . . . so as not to forget to work on them. Fifteen years later, I come into his office . . . and we're talking about something, and I look up at his bulletin board and [notice that] that sheet of paper is still there but the sun has faded everything that was written on it. So the problems went away!"

To be fair, we've made two experimental discoveries in the past few decades: that neutrinos have mass and that the universe is dominated by a mysterious dark energy that seems to be accelerating its expansion. But we have no idea *why* neutrinos (or any of the other particles) have mass or what explains their mass value. As for the dark energy, it's not explained in terms of any existing theory. Its discovery cannot then be counted as a success, for it suggests that there is some major fact we are all missing. And except for the dark energy, no new particle has been discovered, no new force found, no new phenomenon encountered that was not known and understood twenty-five years ago.

Don't get me wrong. For the past twenty-five years we have certainly been very busy. There has been enormous progress in applying established theories to diverse subjects: the properties of materials, the molecular physics underlying biology, the dynamics of vast clusters of stars. But when it comes to extending our knowledge of the laws of nature, we have made no real headway. Many beautiful ideas have been explored, and there have been remarkable particle-accelerator experiments and cosmological observations, but these have mainly served to confirm existing theory. There have been few leaps forward, and none as definitive or important as those of the previous two hundred years. When something like this happens in sports or business, it's called hitting the wall.

Why is physics suddenly in trouble? And what can we do about it? These are the central questions of my book.

I'm an optimist by nature, and for a long time I fought the conclusion that this period in physics — the period of my own career — has been an unusually fallow one. For me and many of my friends who entered science with the hope of making important contributions to what then was a rapidly moving field, there is a shocking fact we must come to terms with: Unlike any previous generation, we have not achieved anything that we can be confident will outlive us. This has given rise to personal crises. But, more important, it has produced a crisis in physics.

The main challenge for theoretical particle physics over the last three decades has been to explain the standard model more deeply. Here there has been a lot of activity. New theories have been posited and explored, some in great detail, but none has been confirmed experimentally. And here's the crux of the problem: In science, for a theory to be believed, it must make a new prediction — different from those made by previous theories — for an experiment not yet done. For the experiment to be meaningful, we must be able to get an answer that disagrees with that prediction. When this is the case, we say that a theory is *falsifiable* — vulnerable to being shown false. The theory also has to be *confirmable*; it must be possible to verify a new prediction that only this theory makes. Only when a theory has been tested and the results agree with the theory do we advance the theory to the ranks of true theories.

The current crisis in particle physics springs from the fact that the theories that have gone beyond the standard model in the last thirty years fall into two categories. Some were falsifiable, and they were falsified. The rest are untested — either because they make no clean predictions or because the predictions they do make are not testable with current technology.

Over the last three decades, theorists have proposed at least a dozen new approaches. Each approach is motivated by a compelling hypothesis, but none has so far succeeded. In the realm of particle physics, these include Technicolor, preon models, and supersymmetry. In the realm of spacetime, they include twistor theory, causal sets, supergravity, dynamical triangulations, and loop quantum gravity. Some of these ideas are as exotic as they sound.

One theory has attracted more attention than all the others combined: string theory. The reasons for its popularity are not hard to understand. It purports to correctly describe the big and the small — both gravity and the elementary particles — and to do so, it makes the boldest hypotheses of all the theories: It posits that the world contains as yet unseen dimensions and many more particles than are presently known. At the same time, it proposes that all the elementary particles arise from the vibrations of a single entity — a string — that obeys simple and beautiful laws. It claims to be the one theory that unifies *all* the particles and *all* the forces in nature. As such, it promises to make clean and unambiguous predictions for any experiment that has ever been done or ever could be done. Much effort has been put into string theory in the last twenty years, but we still do not know whether it is true. Even after all this work, the theory makes no new predictions that are testable by current — or even currently conceivable — experiments. The few clean predictions it does make have already been made by other well-accepted theories.

Part of the reason string theory makes no new predictions is that it appears to come in an infinite number of versions. Even if we restrict ourselves to theories that agree with some basic observed facts about our universe, such as its vast size and the existence of the dark energy, we are left with as many as 10^{500} distinct string theories — that's 1 with 500 zeros after it, more than all the atoms in the known universe. With such a vast number of theories, there is little hope that we can identify an outcome of an experiment that would not be encompassed by one of them. Thus, no matter what the experiments show, string theory cannot be disproved. But the reverse also holds: No experiment will ever be able to prove it true.

At the same time, we understand very little about most of these string theories. And of the small number we do understand in any detail, every single one disagrees with the present experimental data, usually in at least two ways.

So we face a paradox. Those string theories we know how to study are known to be wrong. Those we cannot study are thought to exist in such vast numbers that no conceivable experiment could ever disagree with all of them.

These are not the only problems. String theory rests on several key conjectures, for which there is some evidence but no proof. Even worse, after all the scientific labor expended in its study, we still do not know whether there is a complete and coherent theory that can even go by the name "string theory." What we have, in fact, is not a theory at all but a large collection of approximate calculations, together with a web of conjectures that, if true, point to the existence of a theory. But that theory has never actually been written down. We don't know what its fundamental principles are. We don't know what mathematical language it should be expressed in — perhaps a new one will have to be invented to describe it. Lacking both fundamental principles and the mathematical formulation, we cannot say that we even know what string theory asserts.

Here is how the string theorist Brian Greene puts it in his latest book, *The Fabric of the Cosmos:* "Even today, more than three decades after its initial articulation, most string practitioners believe we still don't have a comprehensive answer to the rudimentary question, What is string theory? . . . [M]ost researchers feel that our current formulation of string theory still lacks the kind of core principle we find at the heart of other major advances."[2]

Gerard 't Hooft, a Nobel Prize winner for his work in elementary-particle physics, has characterized the state of string theory this way: "Actually, I would not even be prepared to call string theory a 'theory,' rather a 'model,' or not even that: just a hunch. After all, a theory should come with instructions on how to deal with it to identify the things one wishes to describe, in our case the elementary particles, and one should, at least in principle, be able to formulate the rules for calculating the properties of these particles, and how to make new predictions for them. Imagine that I give you a chair, while explaining that the legs are still missing, and that the seat, back and armrest will perhaps be delivered soon. Whatever I did give you, can I still call it a chair?"[3]

David Gross, a Nobel laureate for his work on the standard model, has since become one of the most aggressive and formidable champions of string theory. Yet he closed a recent conference intended to celebrate the theory's progress by saying, "We don't know what we are talking about. . . . The state of physics today is like it was when we were mystified by radioactivity. . . . They were missing something absolutely fundamental. We are missing perhaps something as profound as they were back then."[4]

But though string theory is so incomplete that its very existence is an unproved conjecture, that does not keep many who work on it from believing that it is the only way forward for theoretical physics. One prominent string theorist, Joseph Polchinski, of the Kavli Institute for Theoretical Physics at UC Santa Barbara, was asked not long ago to give a talk on "Alternatives to String Theory." His first reaction, he said, "was that this was silly, there are no alternatives. . . . All good ideas are part of string theory."[5] Lubos Motl, an assistant professor at Harvard, recently asserted on his blog that "the most likely reason why no . . . person has convinced others about [an] alternative to string theory is that there probably exists no alternative to string theory."[6]

What is going on here? Usually in science one means something quite definite by the term *theory*. Lisa Randall, an influential particle theorist and Motl's colleague at Harvard, defines a theory as "a definite physical framework embodied in a set of fundamental assumptions about the world — and an economical framework that encompasses a wide variety of phenomena. A theory yields a specific set of equations and predictions — ones that are borne out by successful agreement with experimental results."[7]

String theory does not fit this description — at least not yet. How, then, are some experts sure there is no alternative to string theory, if they don't know precisely what it is? What exactly is it that they are sure has no alternative? These are some of the questions that led me to write this book.

Theoretical physics is hard. Very hard. Not because a certain amount of math is involved but because it involves great risks. As we will see over and over again as we examine the story of contemporary physics, science of this kind cannot be done without risk. If a large number of people have worked on a question for many years and the answer remains unknown, it may mean that the answer is not easy or obvious. Or this may be a question that has no answer.

String theory, to the extent it is understood, posits that the world is fundamentally different from the world we know. If string theory is right, the world has more dimensions and many more particles and forces than we have so far observed. Many string theorists talk and write as if the existence of those extra dimensions and particles were an assured fact, one that no good scientist can doubt. More than once, a string theorist has said to me something like "But do you mean you think it's possible that there are *not* extra dimensions?" In fact, neither theory nor experiment offers any evidence at all that extra dimensions exist. One of the goals of this book is to demystify the claims of string theory. The ideas are beautiful and well motivated. But to understand why they have not led to greater progress, we have to be clear about exactly what the evidence supports and what is still missing.

Because string theory is such a high-risk venture — unsupported by experiment, though very generously supported by the academic and scientific communities — there are only two ways the story can end. If string theory turns out to be right, string theorists will turn out to be the greatest heroes in the history of science. On the basis of a handful of clues — none of which has an unambiguous reading — they will have discovered that reality is far more vast than previously imagined. Columbus discovered a new continent unknown to the king and queen of Spain (as the Spanish royals were unknown to the residents of the New World). Galileo discovered new stars and moons, and later astronomers discovered new planets. All this would pale in the face of the discovery of new dimensions. Moreover, many string theorists believe that the myriad worlds described by the huge number of string theories really do exist — as other universes impossible for us to see directly. If they are right, we see far less of reality than any group of cave dwellers saw of the earth. No one in human history has ever guessed correctly about such a large expansion of the known world.

On the other hand, if string theorists are wrong, they can't be just a little wrong. If the new dimensions and symmetries do not exist, then we will count string theorists among science's greatest failures, like those who continued to work on Ptolemaic epicycles while Kepler and Galileo forged ahead. Theirs will be a cautionary tale of how not to do science, how not to let theoretical conjecture get so far beyond the limits of what can rationally be argued that one starts engaging in fantasy.

One result of the rise of string theory is that the community of people who work on fundamental physics is split. Many scientists continue to work on string theory, and perhaps as many as fifty new PhDs are awarded each year for work in this field. But there are some physicists who are deeply skeptical — who either never saw the point or have by now given up waiting for a sign that the theory has a consistent formulation or makes a real experimental prediction. The split is not always friendly. Doubts are expressed on each side about the professional competence and ethical standards of the other, and it is real work maintaining friendships across the divide.

According to the picture of science we all learned in school, situations like this are not supposed to develop. The whole point of modern science, we are taught, is that there is a method that leads to progress in our understanding of nature. Disagreement and controversy are of course necessary for science to progress, but there is always supposed to be a way to resolve a dispute by means of experiment or mathematics. In the case of string theory, however, this mechanism seems to have broken down. Many adherents and critics of string theory are so confirmed in their views that it is difficult to have a cordial discussion on the issue, even among friends. "How can you not see the beauty of the theory? How could a theory do all this and not be true?" say the string theorists. This provokes an equally heated response from skeptics: "Have you lost your mind? How can you believe so strongly in *any* theory in the complete absence of experimental test? Have you forgotten how science is supposed to work? How can you be so sure you are right when you do not even know what the theory is?"

I have written this book in the hope that it will contribute to an honest and useful discussion among experts and lay readers alike. In spite of what I have seen in the last few years, I believe in science. I believe in the ability of the scientific community to rise above acrimony and resolve controversy through rational argument based on the evidence in front of us. I am aware that just by raising these issues, I will anger some of my friends and colleagues who work on string theory. I can only insist that I am writing this book not to attack string theory or those who believe in it but out of admiration for them and, above all, as an expression of faith in the physics scientific community.

So this is not a book about "us" versus "them." During my career, I have worked on both string theory and on other approaches to quantum gravity (the reconciliation of Einstein's general theory of relativity with quantum theory). Even if most of my efforts have gone into these other approaches, there have been periods when I avidly believed in string theory and devoted myself to solving its key problems. While I didn't solve them, I wrote eighteen papers in the subject; thus, the mistakes I will discuss are my mistakes as much as anyone else's. I will speak of conjectures that were widely believed to be true, in spite of never having been proved. But I was among the believers, and I made choices about my research based on those beliefs. I will speak of the pressures that young scientists feel to pursue topics sanctioned by the mainstream in order to have a decent career. I have felt those pressures myself, and there were times when I let my career be guided by them. The conflict between the need to make scientific judgments independently and make them in a way that doesn't alienate you from the mainstream is one that I, too, have experienced. I write this book not to criticize scientists who have made choices different from mine but to examine why scientists need to be confronted with such choices at all.

In fact, it took me a long time to decide to write this book. I personally dislike conflict and confrontation. After all, in the kind of science we do, anything worth doing is a risk and all that really matters is what our students' students will think worthy of teaching their own students fifty years down the road. I kept hoping someone in the center of string-theory research would write an objective and detailed critique of exactly what has and has not been achieved by the theory. That hasn't happened.

One reason to take these issues public goes back to the debate that took place a few years ago between scientists and "social constructivists," a group of humanities and social science professors, over how science works. The social constructivists claimed that the scientific community is no more rational or objective than any other community of human beings. This is not how most scientists view science. We tell our students that belief in a scientific theory must always be based on an objective evaluation of the evidence. Our opponents in the debate argued that our claims about how science works were mainly propaganda designed to intimidate people into giving us power, and that the whole scientific enterprise was driven by the same political and sociological forces that drove people in other fields.

One of the main arguments we scientists used in that debate was that our community was different because we governed ourselves according to high standards — standards that prevented us from embracing any theory until it had been proved, by means of published calculations and experimental data, beyond the doubt of a competent professional. As I will relate in some detail, this is not always the case in string theory. Despite the absence of experimental support and precise formulation, the theory is believed by some of its adherents with a certainty that seems emotional rather than rational.

The aggressive promotion of string theory has led to its becoming the primary avenue for exploring the big questions in physics. Nearly every particle theorist with a permanent position at the prestigious Institute for Advanced Study, including the director, is a string theorist; the exception is a person hired decades ago. The same is true of the Kavli Institute for Theoretical Physics. Eight of the nine MacArthur Fellowships awarded to particle physicists since the beginning of the program in 1981 have also gone to string theorists. And in the country's top physics departments (Berkeley, Caltech, Harvard, MIT, Princeton, and Stanford), twenty out of the twenty-two tenured professors in particle physics who received PhDs after 1981 made their reputation in string theory or related approaches.

String theory now has such a dominant position in the academy that it is practically career suicide for young theoretical physicists not to join the field. Even in areas where string theory makes no predictions, like cosmology and particle phenomenology, it is common for researchers to begin talks and papers by asserting a belief that their work will be derivable from string theory sometime in the future.

There are good reasons to take string theory seriously as a hypothesis about nature, but this is not the same as declaring its truth. I invested several years of work in string theory because I believed in it enough to want to try my hand at solving its key problems. I also believed that I had no right to an opinion until I knew it in detail, as only a practitioner could. At the same time, I have worked on other approaches that also promise to answer fundamental questions. As a result, I'm regarded with some suspicion by people on both sides of the debate. Some string theorists consider me "anti-string." This couldn't be less true. I would never have put so much time and effort into working on string theory, or written three books largely motivated by its problems, if I wasn't fascinated by it and didn't feel that it might turn out to be part of the truth. Nor am I *for* anything except science, or *against* anything except that which threatens science.

But there's more at stake than amity among colleagues. To do our work, we physicists require significant resources, which are provided largely by our fellow citizens — through taxes as well as foundation money. In exchange, they ask only for the chance to look over our shoulders as we forge ahead and deepen humanity's knowledge of the world we share. Those physicists who communicate with the public, whether through writing, public speaking, television, or the Internet, have a responsibility to tell the story straight. We must be careful to present the failures along with the successes. Indeed, being honest about failures is likely to help rather than hurt our cause. After all, the people who support us live in the real world. They know that progress in any endeavor requires that real risks be taken, that sometimes you will fail.

In recent years, many books and magazine articles for the general public have described the amazing new ideas that theoretical physicists have been working on. Some of these chronicles have been less than careful about explaining just how far the new ideas are from both experimental test and mathematical proof. Having benefited from the public's desire to know how the universe works, I feel a responsibility to make sure that the story told in this book sticks close to the facts. I hope to lay out the various problems we have been unable to solve, explain clearly what experiment supports and doesn't support, and distinguish fact from speculation and intellectual fad.

Above all, we physicists have a responsibility to the future of our craft. Science, as I shall argue later, is based on an ethic, and that ethic requires good faith on the part of its practitioners. It also requires that each scientist be the judge of what he or she believes, so that every unproved idea is met with a healthy dose of skepticism and criticism until it is proved. This, in turn, requires that a diversity of approaches to unsolved problems be supported and welcomed into the community of science. We do research because even the smartest among us doesn't know the answer. Often it lies in a direction other than the one pursued by the mainstream. In those cases, and even when the mainstream guesses right, the progress of science depends on healthy support for scientists who hold divergent views.

Science requires a delicate balance between conformity and variety. Because it is so easy to fool ourselves, because the answers are unknown, experts, no matter how well trained or smart, will disagree about which approach is most likely to yield fruit. Therefore, if science is to move forward, the scientific community must support a variety of approaches to any one problem.

There is ample evidence that these basic principles are no longer being followed in the case of fundamental physics. While few would disagree with the rhetoric of diverse views, it is being practiced less and less. Some young string theorists have told me that they feel constrained to work on string theory whether or not they believe in it, because it is perceived as the ticket to a professorship at a university. And they are right: In the United States, theorists who pursue approaches to fundamental physics other than string theory have almost no career opportunities. In the last fifteen years, there have been a total of three assistant professors appointed to American research universities who work on approaches to quantum gravity other than string theory, and these appointments were all to a single research group. Even as string theory struggles on the scientific side, it has triumphed within the academy.

This hurts science, because it chokes off the investigation of alternative directions, some of them very promising. Despite the inadequate investment in these approaches, a few have moved ahead of string theory to the point of suggesting definite predictions for experiments, which are now in progress.

How is it possible that string theory, which has been pursued by more than a thousand of the brightest and best-educated scientists, working in the best conditions, is in danger of failing? This has puzzled me for a long time, but now I think I know the answer. What I believe is failing is not so much a particular theory but a style of doing science that was well suited to the problems we faced in the middle part of the twentieth century but is ill suited to the kinds of fundamental problems we face now. The standard model of particle physics was the triumph of a particular way of doing science that came to dominate physics in the 1940s. This style is pragmatic and hard-nosed and favors virtuosity in calculating over reflection on hard conceptual problems. This is profoundly different from the way that Albert Einstein, Niels Bohr, Werner Heisenberg, Erwin Schrödinger, and the other early-twentieth-century revolutionaries did science. Their work arose from deep thought on the most basic questions surrounding space, time, and matter, and they saw what they did as part of a broader philosophical tradition, in which they were at home.

In the approach to particle physics developed and taught by Richard Feynman, Freeman Dyson, and others, reflection on foundational problems had no place in research. This freed them from the debates over the meaning of quantum physics that their elders were embroiled in and led to thirty years of dramatic progress. This is as it should be: Different styles of research are needed to solve different kinds of problems. Working out the applications of established frameworks requires very different kinds of thinking — and thinkers — than inventing those frameworks in the first place.

However, as I will argue in detail in the pages to come, the lesson of the last thirty years is that the problems we're up against today cannot be solved by this pragmatic way of doing science. To continue the progress of science, we have to again confront deep questions about space and time, quantum theory, and cosmology. We again need the kinds of people who can invent new solutions to long-standing foundational problems. As we shall see, the directions in which progress is being made — which are taking theory back into contact with experiment — are led by people who have an easier time inventing new ideas than following popular trends and for the most part do science in the reflective and foundational style of the early-twentieth-century pioneers.

I want to emphasize that my concern is not with string theorists as individuals, some of whom are the most talented and accomplished physicists I know. I would be the first to defend their right to pursue the research they think is most promising. But I am extremely concerned about a trend in which only one direction of research is well supported while other promising approaches are starved.

It is a trend with tragic consequences if, as I will argue, the truth lies in a direction that requires a radical rethinking of our basic ideas about space, time, and the quantum world.

I

THE UNFINISHED
REVOLUTION

1

The Five Great Problems
in Theoretical Physics

FROM THE BEGINNING of physics, there have been those who imagined they would be the last generation to face the unknown. Physics has always seemed to its practitioners to be almost complete. This complacency is shattered only during revolutions, when honest people are forced to admit that they don't know the basics. But even revolutionaries still imagine that the big idea — the one that will tie it all up and end the search for knowledge — lies just around the corner.

We live in one of those revolutionary periods, and have for a century. The last such period was the Copernican revolution, beginning in the early sixteenth century, during which Aristotelian theories of space, time, motion, and cosmology were overthrown. The culmination of that revolution was Isaac Newton's proposal of a new theory of physics, published in his *Philosophiae Naturalis Principia Mathematica* in 1687. The current revolution in physics began in 1900, with Max Planck's discovery of a formula describing the energy distribution in the spectrum of heat radiation, which demonstrated that the energy is not continuous but quantized. This revolution has yet to end. The problems that physicists must solve today are, to a large extent, questions that remain unanswered because of the incompleteness of the twentieth century's scientific revolution.

The core of our failure to complete the present scientific revolution consists of five problems, each famously intractable. These problems confronted us when I began my study of physics in the 1970s, and while we have learned a lot about them in the last three decades, they remain unsolved. One way or another, any proposed theory of fundamental physics must solve these five problems, so it's worth taking a closer look at each.

Albert Einstein was certainly the most important physicist of the twentieth century. Perhaps his greatest work was his discovery of general relativity, which is the best theory we have so far of space, time, motion, and gravitation. His profound insight was that gravity and motion are intimately related to each other and to the geometry of space and time. This idea broke with hundreds of years of · thinking about the nature of space and time, which until then had been viewed as fixed and absolute. Being eternal and unchanging, they provided a background, which we used to define notions like position and energy.

In Einstein's general theory of relativity, space and time no longer provide a fixed, absolute background. Space is as dynamic as matter; it moves and morphs. As a result, the whole universe can expand or shrink, and time can even begin (in a Big Bang) and end (in a black hole).

Einstein accomplished something else as well. He was the first person to understand the need for a new theory of matter and radiation. Actually, the need for a break was implicit in Planck's formula, but Planck had not understood its implications deeply enough; he felt that it could be reconciled with Newtonian physics. Einstein thought otherwise, and he gave the first definitive argument for such a theory in 1905. It took twenty more years to invent that theory, known as the quantum theory.

These two discoveries, of relativity and of the quantum, each required us to break definitively with Newtonian physics. However, in spite of great progress over the century, they remain incomplete. Each has defects that point to the existence of a deeper theory. But the main reason each is incomplete is the existence of the other.

The mind calls out for a third theory to unify all of physics, and for a simple reason. Nature is in an obvious sense "unified." The universe we find ourselves in is interconnected, in that everything interacts with everything else. There is no way we can have two theories of nature covering different phenomena, as if one had nothing to do with the other. Any claim for a final theory must be a complete theory of nature. It must encompass all we know.

Physics has survived a long time without that unified theory. The reason is that, as far as experiment is concerned, we have been able to divide the world into two realms. In the atomic realm, where quantum physics reigns, we can usually ignore gravity. We can treat space and time much as Newton did — as an unchanging background. The other realm is that of gravitation and cosmology. In that world, we can often ignore quantum phenomena.

But this cannot be anything other than a temporary, provisional solution. To go beyond it is the first great unsolved problem in theoretical physics:

Problem 1: Combine general relativity and quantum theory into a single theory that can claim to be the complete theory of nature.

This is called the *problem of quantum gravity.*

Besides the argument based on the unity of nature, there are problems specific to each theory that call for unification with the other. Each has a problem of infinities. In nature, we have yet to encounter anything measurable that has an infinite value. But in both quantum theory and general relativity, we encounter predictions of physically sensible quantities becoming infinite. This is likely the way that nature punishes impudent theorists who dare to break her unity.

General relativity has a problem with infinities because inside a black hole the density of matter and the strength of the gravitational field quickly become infinite. That appears to have also been the case very early in the history of the universe — at least, if we trust general relativity to describe its infancy. At the point at which the density becomes infinite, the equations of general relativity break down. Some people interpret this as time stopping, but a more sober view is that the theory is just inadequate. For a long time, wise people have speculated that it is inadequate because the effects of quantum physics have been neglected.

Quantum theory, in turn, has its own trouble with infinities. They appear whenever you attempt to use quantum mechanics to describe fields, like the electromagnetic field. The problem is that the electric and magnetic fields have values at every point in space.

This means that there are an infinite number of variables (even in a finite volume there are an infinite number of points, hence an infinite number of variables). In quantum theory, there are uncontrollable fluctuations in the values of every quantum variable. An infinite number of variables, fluctuating uncontrollably, can lead to equations that get out of hand and predict infinite numbers when you ask questions about the probability of some event happening, or the strength of some force.

So this is another case where we can't help but feel that an essential part of physics has been left out. There has long been the hope that when gravity is taken into account, the fluctuations will be tamed and all will be finite. If infinities are signs of missing unification, a unified theory will have none. It will be what we call a *finite theory,* a theory that answers every question in terms of sensible, finite numbers.

Quantum mechanics has been extremely successful at explaining a vast realm of phenomena. Its domain extends from radiation to the properties of transistors and from elementary-particle physics to the action of enzymes and other large molecules that are the building blocks of life. Its predictions have been borne out again and again over the course of the last century. But some physicists have always had misgivings about it, because the reality it describes is so bizarre. Quantum theory contains within it some apparent conceptual paradoxes that even after eighty years remain unresolved. An electron appears to be both a wave and a particle. So does light. Moreover, the theory gives only statistical predictions of subatomic behavior. Our ability to do any better than that is limited by the *uncertainty principle,* which tells us that we cannot measure a particle's positon and momentum at the same time. The theory yields only probabilities. A particle — an atomic electron, say — can be anywhere until we measure it; our observation in some sense determines its state. All of this suggests that quantum theory does not tell the whole story. As a result, in spite of its success, there are many experts who are convinced that quantum theory hides something essential about nature that we need to know.

One problem that has bedeviled the theory from the beginning is the question of the relationship between reality and the formalism. Physicists have traditionally expected that science should give an account of reality as it would be in our absence. Physics should be more than a set of formulas that predict what we will observe in an experiment; it should give a picture of what reality *is*. We are accidental descendants of an ancient primate, who appeared only very recently in the history of the world. It cannot be that reality depends on our existence. Nor can the problem of no observers be solved by raising the possibility of alien civilizations, for there was a time when the world existed but was far too hot and dense for organized intelligence to exist.

Philosophers call this view *realism*. It can be summarized by saying that the real world out there (or RWOT, as my first philosophy teacher used to put it) must exist independently of us. It follows that the terms by which science describes reality cannot involve in any essential way what we choose to measure or not measure.

Quantum mechanics, at least in the form it was first proposed, did not fit easily with realism. This is because the theory presupposed a division of nature into two parts. On one side of the division is the system to be observed. We, the observers, are on the other side. With us are the instruments we use to prepare experiments and take measurements, and the clocks we use to record when things happen. Quantum theory can be described as a new kind of language to be used in a dialogue between us and the systems we study with our instruments. This quantum language contains verbs that refer to our preparations and measurements and nouns that refer to what is then seen. It tells us nothing about what the world would be like in our absence.

Since quantum theory was first proposed, a debate has raged between those who accept this way of doing science and those who reject it. Many of the founders of quantum mechanics, including Einstein, Erwin Schrödinger, and Louis de Broglie, found this approach to physics repugnant. They were realists. For them quantum theory, no matter how well it worked, was not a complete theory, because it did not provide a picture of reality absent our interaction with it. On the other side were Niels Bohr, Werner Heisenberg, and many others. Rather than being appalled, they embraced this new way of doing science.

Since then, the realists have scored some successes by pointing to inconsistencies in the present formulation of quantum theory. Some of these apparent inconsistencies arise because, if it is universal, quantum theory should also describe *us*. Problems, then, come from the division of the world required to make sense of quantum theory. One difficulty is where you draw the dividing line, which depends on who is doing the observing. When you measure an atom, you and your instruments are on one side and the atom is on the other side. But suppose I watch you working through a videocam I have set up in your laboratory. I can consider your whole lab — including you and your instruments, as well as the atoms you play with — to constitute one system that I am observing. On the other side would be only me.

You and I hence describe two different "systems." Yours includes just the atom. Mine includes you, the atom, and everything you use to study it. What you see as a measurement, I see as two physical systems interacting with each other. Thus, even if you agree that it's fine to have the observers' actions as part of the theory, the theory as given is not sufficient. Quantum mechanics has to be expanded, to allow for many different descriptions, depending on who the observer is.

This whole issue goes under the name *the foundational problems of quantum mechanics*. It is the second great problem of contemporary physics.

Problem 2: Resolve the problems in the foundations of quantum mechanics, either by making sense of the theory as it stands or by inventing a new theory that does make sense.

There are several different ways one might do this.

1. Provide a sensible language for the theory, one that resolves all puzzles like the ones just mentioned and incorporates the division of the world into system and observer as an essential feature of the theory.
2. Find a new interpretation of the theory — a new way of reading the equations — that is realist, so that measurement and observation play no role in the description of fundamental reality.
3. Invent a new theory, one that gives a deeper understanding of nature than quantum mechanics does.

All three options are currently being pursued by a handful of smart people. There are unfortunately not many physicists who work on this problem. This is sometimes taken as an indication that the problem is either solved or unimportant. Neither is true. This is probably the most serious problem facing modern science. It is just so hard that progress is very slow. I deeply admire the physicists who work on it, both for the purity of their intentions and for their courage to ignore fashion and attack the hardest and most fundamental of problems.

But despite their best efforts, the problem remains unsolved. This suggests to me that it's not just a matter of finding a new way to think about quantum theory. Those who initially formulated the theory were not realists. They did not believe that human beings were capable of forming a true picture of the world as it exists independent of our actions and observations. They argued instead for a very different vision of science: In their view, science can be nothing but an extension of the ordinary language we use to describe our actions and observations to one another.

In more recent times, that view looks self-indulgent — the product of a time we hope we have advanced beyond in many respects. Those who continue to defend quantum mechanics as formulated, and propose it as a theory of the world, do so mostly under the banner of realism. They argue for a reinterpretation of the theory along realist lines. However, while they have made some interesting proposals, none has been totally convincing.

It is possible that realism as a philosophy will simply die off, but this seems unlikely. After all, realism provides the motivation driving most scientists. For most of us, belief in the RWOT and the possibility of truly knowing it motivates us to do the hard work needed to become a scientist and contribute to the understanding of nature. Given the failure of realists to make sense of quantum theory as formulated, it appears more and more likely that the only option is the third one: the discovery of a new theory that will be more amenable to a realist interpretation.

I should admit that I am a realist. I side with Einstein and the others who believe that quantum mechanics is an incomplete description of reality. Where, then, should we look for what is missing in quantum mechanics? It has always seemed to me that the solution will require more than a deeper understanding of quantum physics itself. I believe that if the problem has not been solved after all this time, it is because there is something missing, some link to other problems in physics. The problem of quantum mechanics is unlikely to be solved in isolation; instead, the solution will probably emerge as we make progress on the greater effort to unify physics.

But if this is true, it works both ways: We will not be able to solve the other big problems unless we also find a sensible replacement for quantum mechanics.

The idea that physics should be unified has probably motivated more work in physics than any other problem. But there are different ways that physics can be unified, and we should be careful to distinguish them. So far we have been discussing *unification through a single law*. It is hard to see how anyone could disagree that this is a necessary goal.

But there are other ways to unify the world. Einstein, who certainly thought as much about this as anyone, emphasized that we must distinguish two kinds of theories. There are *theories of principle* and *constructive theories*. A theory of principle is one that sets up the framework that makes a description of nature possible. By definition, a theory of principle must be universal: It must apply to everything because it sets out the basic language we use to talk about nature. There cannot be two different theories of principle, applying to different domains. Because the world is a unity, everything interacts ultimately with everything else, and there can be only one language used to describe those interactions. Quantum theory and general relativity are both theories of principle. As such, logic requires their unification.

The other kind of theories, constructive theories, describe some particular phenomenon in terms of specific models or equations.[1] The theory of the electromagnetic field and the theory of the electron are constructive theories. Such a theory cannot stand alone; it must be set within the context of a theory of principle. But as long as the theory of principle allows, there can be phenomena that obey different laws. For example, the electromagnetic field obeys laws different from those governing the postulated cosmological dark matter (thought to vastly outnumber the amount of ordinary atomic matter in our universe). One thing we know about the dark matter is that, whatever it is, it is dark. This means it gives off no light, so it likely doesn't interact with the electromagnetic field. Thus two different theories can coexist side by side.

The point is that the laws of electromagnetism do not dictate what else exists in the world. There can be quarks or not, neutrinos or not, dark matter or not. Similarly, the laws that describe the two forces — strong and weak — that act within the atomic nucleus do not necessarily require that there be an electromagnetic force. We can easily imagine a world with electromagnetism but no strong nuclear force, or the reverse. As far as we know, either possibility would be consistent.

But it is still possible to ask whether all the forces we observe in nature might be manifestations of a single, fundamental force. There seems, as far as I can tell, no logical argument that this *should* be true, but it is still something that *might* be true.

The desire to unify the various forces has led to several significant advances in the history of physics. James Clerk Maxwell, in 1867, unified electricity and magnetism into one theory, and a century later, physicists realized that the electromagnetic field and the field that propagates the weak nuclear force (the force responsible for radioactive decay) could be unified. This became the *electroweak* theory, whose predictions have been repeatedly confirmed in experiments over the last thirty years.

There are two fundamental forces in nature (that we know of) that remain outside the unification of the electromagnetic and weak fields. These are gravity and the strong nuclear force, the force responsible for binding the particles called quarks together to form the protons and neutrons making up the atomic nucleus. Can all four fundamental forces be unified?

This is our third great problem.

Problem 3: Determine whether or not the various particles and forces can be unified in a theory that explains them all as manifestations of a single, fundamental entity.

Let us call this problem *the unification of the particles and forces*, to distinguish it from the unification of laws, the unification we discussed earlier.

At first, this problem appears easy. The first proposal for how to unify gravity with electricity and magnetism was made in 1914, and many more have been offered since. They all work, as long as you forget one thing, which is that nature is quantum mechanical. If you leave quantum physics out of the picture, unified theories are easy to invent. But if you include quantum theory, the problem gets much, much harder. Since gravity is one of the four fundamental forces of nature, we must solve the problem of quantum gravity (that is, problem no. 1: how to reconcile general relativity and quantum theory) along with the problem of unification.

Over the last century, our physical description of the world has simplified quite a bit. As far as particles are concerned, there appear to be only two kinds, quarks and leptons. Quarks are the constituents of protons and neutrons and many particles we have discovered similar to them. The class of leptons encompasses all particles not made of quarks, including electrons and neutrinos. Altogether, the known world is explained by six kinds of quarks and six kinds of leptons, which interact with each other through the four forces (or interactions, as they are also known): gravity, electromagnetism, and the strong and weak nuclear forces.

Twelve particles and four forces are all we need to explain everything in the known world. We also understand very well the basic physics of these particles and forces. This understanding is expressed in terms of a theory that accounts for all of these particles and all of the forces except for gravity. It's called *the standard model of elementary-particle physics* — or the standard model, for short. This theory does not have the problem of infinities mentioned earlier. Anything we want to compute in this theory we can, and it results in a finite number. In the more than thirty years since it was formulated, many predictions made by this theory have been checked experimentally. In each and every case, the theory has been confirmed.

The standard model was formulated in the early 1970s. Except for the discovery that neutrinos have mass, it has not required adjustment since. So why wasn't physics over by 1975? What remained to be done?

For all its usefulness, the standard model has a big problem: It has a long list of adjustable constants. When we state the laws of the theory, we must specify the values of these constants. As far as we know, any values will do, because the theory is mathematically consistent no matter which values we put in. These constants specify the properties of the particles. Some tell us the masses of the quarks and the leptons, while others tell us the strengths of the forces. We have no idea why these numbers have the values they do; we simply determine them by experiments and then plug in the numbers. If you think of the standard model as a calculator, then the constants will be dials that can be set to whatever positions you like each time the program is run.

There are about twenty such constants, and the fact that there are that many freely specifiable constants in what is supposed to be a fundamental theory is a tremendous embarrassment. Each one represents some basic fact of which we are ignorant: namely, the physical reason or mechanism responsible for setting the constant to its observed value.

This is our fourth big problem.

Problem 4: Explain how the values of the free constants in the standard model of particle physics are chosen in nature.

It is devoutly hoped that a true unified theory of the particles and forces will give a unique answer to this question.

In 1900, William Thomson (Lord Kelvin), an influential British physicist, famously proclaimed that physics was over, except for two small clouds on the horizon. These "clouds" turned out to be the clues that led us to quantum theory and relativity theory. Now, even as we celebrate the encompassing of all known phenomena in the standard model plus general relativity, we, too, are aware of two clouds. These are the dark matter and the dark energy.

Apart from the issue of its relationship with the quantum, we think we understand gravity very well. The predictions of general relativity have been found to be in agreement with observation to a very precise degree. The observations in question extend from falling bodies and light on Earth, to the detailed motion of the planets and their moons, to the scales of galaxies and clusters of galaxies. Formerly exotic phenomena — such as gravitational lensing, an effect of the curvature of space by matter — are now so well understood that they are used to measure the distributions of mass in galactic clusters.

In many cases — those in which velocities are small compared with that of light, and masses are not too compact — Newton's laws of gravity and motion provide an excellent approximation to the predictions of general relativity. Certainly they should help us predict how the motion of a particular star is influenced by the masses of stars and other matter in its galaxy. But they don't. Newton's law of gravity says that the acceleration of any object as it orbits another is proportional to the mass of the body it is orbiting. The heavier the star, the faster the orbital motion of the planet. That is, if two stars are each orbited by a planet, and the planets are the same distances from their stars, the planet orbiting the more massive star will move faster. Thus if you know the speed of a body in orbit around a star and its distance from the star, you can measure the mass of that star. The same holds for stars in orbit around the center of their galaxy; by measuring the orbital speeds of the stars, you can measure the distribution of mass in that galaxy.

Over the last decades, astronomers have done a very simple experiment in which they measure the distribution of mass in a galaxy in two different ways and compare the results. First, they measure the mass by observing the orbital speeds of the stars; second, they make a more direct measurement of the mass by counting all the stars, gas, and dust they can see in the galaxy. The idea is to compare the two measurements: Each should tell them both the total mass in the galaxy and how it is distributed. Given that we understand gravity well, and that all known forms of matter give off light, the two methods should agree.

They don't. Astronomers have compared the two methods of measuring mass in more than a hundred galaxies. In almost all cases, the two measurements don't agree, and not by just a small amount but by factors of up to 10. Moreover, the error always goes in one direction: There is always more mass needed to explain the observed motions of the stars than is seen by directly counting up all the stars, gas, and dust.

There are only two explanations for this. Either the second method fails because there is much more mass in a galaxy than is visible, or Newton's laws fail to correctly predict the motions of stars in the gravitational field of their galaxy.

All the forms of matter we know about give off light, either directly as in starlight or reflected from planets or interstellar rocks, gas, and dust. So if there is matter we don't see, it must be in some novel form that neither emits nor reflects light. And because the discrepancy is so large, the majority of the matter in galaxies must be in this new form.

Today most astronomers and physicists believe that this is the right answer to the puzzle. There is missing matter, which is actually there but which we don't see. This mysterious missing matter is referred to as the *dark matter.* The dark-matter hypothesis is preferred mostly because the only other possibility — that we are wrong about Newton's laws, and by extension general relativity — is too scary to contemplate.

Things have become even more mysterious. We have recently discovered that when we make observations at still larger scales, corresponding to billions of light-years, the equations of general relativity are not satisfied even when the dark matter is added in. The expansion of the universe, set in motion by the Big Bang some 13.7 billion years ago, appears to be accelerating, whereas, given the observed matter plus the calculated amount of dark matter, it should be doing the opposite — decelerating.

Again, there are two possible explanations. General relativity could simply be wrong. It has been verified precisely only within our solar system and nearby systems in our own galaxy. Perhaps when one gets to a scale comparable to the size of the whole universe, general relativity is simply no longer applicable.

Or there is a new form of matter — or energy (recall Einstein's famous equation $E = mc^2$, showing the equivalence of energy and mass) — that becomes relevant on these very large scales: That is, this new form of energy affects only the expansion of the universe. To do this, it cannot clump around galaxies or even clusters of galaxies. This strange new energy, which we have postulated to fit the data, is called the *dark energy.*

Most kinds of matter are under pressure, but the dark energy is under tension — that is, it pulls things together rather than pushes them apart. For this reason, tension is sometimes called negative pressure. In spite of the fact that the dark energy is under tension, it causes the universe to expand faster. If you are confused by this, I sympathize. One would think that a gas with negative pressure would act like a rubber band connecting the galaxies and slow the expansion down. *But it turns out that when the negative pressure is negative enough, in general relativity it has the opposite effect.* It causes the expansion of the universe to accelerate.

Recent measurements reveal a universe consisting mostly of the unknown. Fully 70 percent of the matter density appears to be in the form of dark energy. Twenty-six percent is dark matter. Only 4 percent is ordinary matter. So less than 1 part in 20 is made out of matter we have observed experimentally or described in the standard model of particle physics. Of the other 96 percent, apart from the properties just mentioned, we know absolutely nothing.

In the last ten years, cosmological measurements have gotten much more precise. This is partly a side effect of Moore's law, which states that every eighteen months or so, the processing speeds of computer chips will double. All the new experiments use microchips in either satellites or ground-based telescopes, so as the chips have gotten better, so have the observations. Today we know a lot about the basic characteristics of the universe, such as the overall matter density and the rate of expansion. There is now a standard model of cosmology, just as there is a standard model of elementary-particle physics. Just like its counterpart, the standard model of cosmology has a list of freely specifiable constants — in this case, about fifteen. These denote, among other things, the density of different kinds of matter and energy and the expansion rate. No one knows anything about why these constants have the values they do. As in particle physics, the values of the constants are taken from observations but are not yet explained by any theory.

These cosmological mysteries make up the fifth great problem.

Problem 5: Explain dark matter and dark energy. Or, if they don't exist, determine how and why gravity is modified on large scales. More generally, explain why the constants of the standard model of cosmology, including the dark energy, have the values they do.

These five problems represent the boundaries to present knowledge. They are what keep theoretical physicists up at night. Together they drive most current work on the frontiers of theoretical physics.

Any theory that claims to be a fundamental theory of nature

must answer each one of them. One of the aims of this book is to evaluate just how well recent physical theories, such as string theory, have done in achieving this goal. But before we do that, we need to examine some earlier attempts at unification. We have a great deal to learn from the successes — and also from the failures.

2

The Beauty Myth

THE MOST CHERISHED goal in physics, as in bad romance novels, is unification. To bring together two things previously understood as different and recognize them as aspects of a single entity — when we can do it — is the biggest thrill in science.

The only sane response to a proposed unification is surprise. *The sun is just another star — and the stars are just suns that happen to be very far away!* Imagine the reaction of a late-sixteenth-century blacksmith or actor on hearing this wild idea of Giordano Bruno's. What could be more absurd than to unify the sun with the stars? People had been taught that the sun was a great fire created by God to warm the earth, while the stars were pinholes in the celestial sphere that let in the light of heaven. Unification instantly turns your world upside down. What you used to believe becomes impossible. *If the stars are suns, the universe is vastly bigger than we thought! Heaven cannot be just overhead!*

Even more important, a new proposal for unification brings with it previously unimagined hypotheses. *If the stars are other suns, there must be planets around them, on which other people live!* The implications often extend beyond science. *If there are other planets with other people on them, then either Jesus came to all of them, in which case his coming to Man was not a unique event, or all those people lose the possibility of salvation!* No wonder the Catholic Church burned Bruno alive.

Great unifications become the founding ideas on which whole new sciences are erected. Sometimes the consequences so threaten our worldview that surprise is quickly followed by disbelief. Before Darwin, each species was in its own eternal category. Each had been made, individually, by God. But evolution by natural selection means that all species have a common ancestor. They are unified into one great family. Biology before Darwin and biology afterward are hardly the same science.

Such powerful new insights lead quickly to new discoveries. *If all living things have a common ancestor, they must be similarly made!* Indeed, we *are* made of the same stuff, because all life turns out to be composed of cells. Plants, animals, fungi, and bacteria seem very different from one another, but they are all just groups of cells arranged in different ways. The chemical processes that construct and power these cells are the same, across the whole empire of life.

If proposals for unification are so shocking to our previous ways of thinking, how is it that people come to believe them? This is in many ways the crux of our story, for it is a story of several proposed unifications, some of which have come to be strongly believed by some scientists. But none of them have achieved consensus among all scientists. As a consequence, we have lively controversy and, at times, emotional debate, the result of the attempted radical alteration of worldviews. So when someone proposes a new unification, how do we tell whether it is true or not?

As you might imagine, not all proposals for unification turn out to be true. At one time, chemists proposed that heat was a substance, like matter. It was called *phlogiston*. This concept unified heat and matter. But it was wrong. The right proposal for the unification of heat and matter is that heat is the energy in random motion of atoms. But although atomism had been proposed by ancient Indian and Greek philosophers, it took until the late nineteenth century before the theory of heat as random motion of atoms was properly developed.

In the history of physics, there have been many proposals for unified theories that turned out to be wrong. A famous one was the idea that light and sound were essentially the same thing: They were both thought to be vibrations in matter. Since sound is vibrations in air, light was proposed to be vibrations in a new kind of matter called the *aether*. Just as the space around us is filled with air, the universe is filled with aether. Einstein killed this particular idea with his own proposal for unification.

All the important ideas that theorists have studied in the last thirty years — such as string theory, supersymmetry, higher dimensions, loops, and others — are proposals for unification. How do we tell which are right and which are not?

I have already mentioned two features that successful unifications tend to share. The first, surprise, cannot be underestimated. If there is no surprise, then the idea is either uninteresting or something we knew before. Second, the consequences must be dramatic: The unification must lead quickly to new insights and hypotheses, becoming the engine that drives progress in understanding.

But there is a third factor that trumps both of these. A good unified theory must offer predictions that no one would have thought to make before. It may even suggest new kinds of experiments that make sense only in light of the new theory. Most important of all, the predictions must be confirmed by experiment.

These three criteria — surprise, new insights, and new predictions confirmed by experiment — are what we will be looking for when we come to judge the promise of current efforts at unification.

Physicists seem to feel a deep need for unification, and some speak as if any step toward further unification must be a step toward the truth. But life is not that simple. At any one time, there can be more than one possible way to unify the things we know — ways that lead science in different directions. In the sixteenth century, there were two very different proposals for unification on the table. There was the old theory, of Aristotle and Ptolemy, according to which the planets were unified with the sun and the moon as part of the celestial spheres. But there was also the new proposal of Copernicus, which unified the planets with Earth. Each had great consequences for science. But at most only one could be right.

We can see here the cost of choosing the wrong unification. If Earth is at the center of the universe, that has tremendous implications for our understanding of motion. In the sky, planets change direction because they are attached to circles whose nature is to rotate eternally. This never happens to things on Earth: Anything we push or throw quickly comes to rest. That is the natural state of things that aren't attached to cosmic circles. Thus in Ptolemy and Aristotle's universe there is a big distinction between being in motion and being at rest.

In their world, there is also a big distinction between the heavens and the earth — things on the earth follow laws different from those that obtain in the sky. Ptolemy proposed that certain bodies in the sky — the sun, the moon, and the five known planets — move on circles that themselves move on circles. These so-called epicycles enabled predictions of eclipses and the motions of the planets — predictions that were accurate to 1 part in 1,000, thus showing the fruitfulness of the unification of sun, moon, and planets. Aristotle gave a natural explanation for Earth's being at the center of the universe: It was composed of Earth-stuff, whose nature was not to move on circles but to seek the center.

To someone educated in that point of view and familiar with how powerfully it explained what we saw around us, Copernicus's proposal that the planets should be considered one with the earth and not the sun must have been profoundly unsettling. If the earth is a planet, then it and everything on it is in continuous motion. How could that be? This violates Aristotle's law that everything not on a celestial circle must come to rest. It also violates experience, for if the earth is moving, how come we don't feel it?

The answer to this puzzle was the greatest unification in all of science: *the unification of motion and rest.* It was proposed by Galileo and codified in Newton's first law of motion, also called *the principle of inertia:* A body at rest or in uniform motion remains in that state of rest or uniform motion unless it is disturbed by forces.

By uniform motion, Newton means motion at a constant speed, in a single direction. Being at rest becomes merely a special case of uniform motion — it is just motion at zero speed.

How can it be that there is no distinction between motion and rest? The key is to realize that whether a body is moving or not has no absolute meaning. Motion is defined only with respect to an observer, who can be moving or not. If you are moving past me at a steady rate, then the cup of coffee I perceive to be at rest on my table is moving with respect to you.

But can't an observer tell whether he is moving or not? To Aristotle, the answer was obviously yes. Galileo and Newton were forced to reply no. If the earth is moving and we do not feel it, then it must be that observers moving at a constant speed do not feel any effect of their motion. Hence we cannot tell whether we are at rest or not, and motion must be defined purely as a relative quantity.

There is an important caveat here: We are talking about uniform motion — motion in a straight line. (While the earth of course doesn't move in a straight line, the deviations from it are too small to feel directly.) When we change the speed or direction of our motion, we do feel it. Such changes are what we call *acceleration,* and acceleration *can* have an absolute meaning.

Galileo and Newton achieved here a subtle and beautiful intellectual triumph. To others, it was obvious that motion and rest were completely different phenomena, easily distinguished. But the principle of inertia unifies them. To explain how it is that they seem different, Galileo invented *the principle of relativity.* This tells us that the distinction between moving and being at rest is meaningful only relative to an observer. Since different observers move differently, they distinguish which objects are moving and which are at rest differently. So the fact that each observer makes a distinction is maintained, as it must be. Thus, whether something is moving or not ceases to be a phenomenon that needs to be explained. For Aristotle, if anything moved, there must be a force acting on it. For Newton, if the motion is uniform, it will persist forever; no force is needed to explain it.

This is a powerful strategy that was repeated in later theories. One way to unify things that appear different is to show that the apparent difference is due to the difference in the perspective of the observers. A distinction that was previously considered absolute becomes relative. This kind of unification is rare and represents the highest form of scientific creativity. When it is achieved, it radically alters our view of the world.

Proposals that two apparently very different things are the same often require a lot of explaining. Only sometimes can you get away with explaining the apparent difference as a consequence of different perspectives. Other times, the two things you choose to unify are just different. The need to then explain how things that seem different are really in some way the same can land a theorist in a lot of trouble.

Let us look at the consequences of Bruno's proposal that the stars are just like our sun. Stars appear much dimmer than the sun. If they are nevertheless the same, then they must be very far away. The distances he had to invoke were much, much larger than the universe was then thought to be. So Bruno's proposal seemed at first absurd.

Of course, this was an opportunity to make a novel prediction: If you could measure the distances to the stars, you would find they were in fact much farther away than the planets. Had it been possible to do this in Bruno's day, he might have escaped the fire. But it was centuries before the distance to a star could be measured. What Bruno had done, in practical terms, was to make an assertion that was untestable, given the technology of the time. Bruno's proposal conveniently put the stars at such a distance that no one could check his idea.

So sometimes the need to explain how things are unified forces you to posit new hypotheses you simply cannot test. This, as we have seen, does not mean you are wrong, but it does mean that originators of new unifications can easily find themselves on dangerous ground.

And it can get worse. Such hypotheses have a habit of compounding themselves. Copernicus, in fact, needed the stars to be very far away. If the stars were as close as Aristotle believed, you could have disproved the motion of the earth — because as the earth moved, the apparent positions of the stars relative to one another would change. To explain why this effect was not seen, Copernicus and his followers had to believe that the stars were very distant. (Of course, we know now that the stars also move, but they are at such tremendous distances that their positions in our sky change extremely slowly.)

But if the stars were so far away, how could we see them? They must be very bright, perhaps as bright as the sun. Hence Bruno's proposal for a universe filled with an infinitude of stars fit naturally with Copernicus's proposal that the earth moved as a planet does.

We see here that different proposals for unification often go together. The proposal that the stars are unified with the sun goes with the proposal that the planets are unified with the earth, and these both require that motion and rest be unified.

These ideas, new in the sixteenth century, opposed another cluster of ideas. Ptolemy's proposal that the planets be unified with the sun and moon and that all move in epicycles went with Aristotle's theory of motion, which unified all known phenomena on the earth.

So we end up with two clusters of ideas, each consisting of several proposals for unification. What is at stake, therefore, is often a whole group of ideas, in which different things are unified at different levels. Before the debate is resolved, there can be good reasons for believing each side. Each side can be supported by observation. Sometimes even the same experiment can be interpreted as evidence for competing theories of unification.

To see how this can happen, consider a ball dropped from the top of a tower. What happens? It falls to the ground and lands at the tower's base. It does not fly off in a westerly direction. Well, you could say, Copernicus and his followers are clearly wrong, for this proves that the earth is not rotating on its axis. Were the earth rotating, the ball would land well away from the base of the tower.

But Galileo and Newton could also claim that the falling ball proves their theory. The principle of inertia tells us that if the ball is moving eastward along with the earth when it is dropped, it will continue to move eastward as it falls. But the ball is moving eastward at the same speed as the tower, so it falls at the tower's base. The same evidence that an Aristotelian philosopher might have used to prove Galileo wrong was taken by Galileo as proof that his theory was correct.

How do we nevertheless decide which proposed unifications are right and which are wrong? At some point, there is a preponderance of evidence. One hypothesis is shown to be so much more fruitful than the other that a rational person has no choice but to agree that the case is proved. With regard to the Newtonian revolution, there was eventually genuine evidence from observation that the earth moved relative to the stars. But before that happened, Newton's laws had proved to be correct in so many instances that there was no going back.

However, in the midst of a scientific revolution there are often rational cases to be made for supporting rival hypotheses. We are in such a period now, and we'll examine conflicting claims for unification in the chapters to come. I will do my best to explain the arguments that support the various sides, while showing why scientists have yet to reach a consensus.

Of course, we do have to exercise caution. Not all evidence said to support a view is solidly based. Sometimes the claims invented to support a theory in trouble are just rationalizations. I recently met a lively group of people standing in the aisle on a flight from London to Toronto. They said hello and asked me where I was coming from, and when I told them I was returning from a cosmology conference, they immediately asked my view on evolution. "Oh no," I thought, then proceeded to tell them that natural selection had been proved true beyond a doubt. They introduced themselves as members of a Bible college on the way back from a mission to Africa, one purpose of which, it turned out, had been to test some of the tenets of creationism. As they sought to engage me in discussion, I warned them that they would lose, as I knew the evidence pretty well. "No," they insisted, "you don't know all the facts." So we got into it. When I said, "But of course you accept the fact that we have fossils of many creatures that no longer live," they responded, "No!"

"What do you mean, 'no'? What about the dinosaurs?"

"The dinosaurs are still alive and roaming the earth!"

"That's ridiculous! Where?"

"In Africa."

"In Africa? Africa is full of people. Dinosaurs are really big. How come no one has seen one?"

"They live deep in the jungle."

"Someone would still have seen one. Do you claim to know someone who has seen one?"

"The pygmies tell us they see them every once in a while. We looked and we didn't see any, but we saw the scratch marks they make eighteen to twenty feet up on the trunks of trees."

"So you agree they are huge animals. And the fossil evidence is that they live in big herds. How could it be that nobody but these pygmies have seen them?"

"That's easy. They spend most of their time hibernating in caves."

"In the jungle? There are caves in the jungle?"

"Yes, of course, why not?"

"Caves big enough for a huge dinosaur to enter? If the caves are so big, they should be easy to find, and you can look inside and see them sleeping."

"To protect themselves while they hibernate, the dinosaurs close up the mouths of their caves with dirt so no one can tell they're there."

"How do they close up the caves so well they can't be seen? Do they use their paws, or perhaps they push the dirt with their noses?"

At this point, the creationists admitted they didn't know, but they told me that "biblical biologists" from their school were in the jungles now, looking for the dinosaurs.

"Be sure to let me know if they bring out a live one," I said, and went back to my seat.

I am not making this up, and I'm not telling this just for your amusement. It illustrates that rationality is not always a simple exercise. Usually it is rational to disbelieve a theory that predicts something that has never been seen. But sometimes there is a good reason for something never having been seen. After all, if there *are* dinosaurs, they must be hiding somewhere. Why not in caves in the African jungle?

This may seem silly, but particle physicists have more than once felt the need to invent an unseen particle, such as the neutrino, in order to make sense of certain theoretical or mathematical results. To explain why it was difficult to detect, they had to make the neutrino interact very weakly. In this case, it was the right strategy, for many years later someone was able to devise an experiment that did find neutrinos. And they did interact very weakly.

So sometimes it is rational not to throw a good theory away when it predicts things that haven't been seen. Sometimes the hypotheses you are forced to invent turn out to be right. By inventing such ad hoc hypotheses, you can not only keep an idea plausible but also sometimes predict new phenomena. But at some point you begin to stretch credulity. The cave-inhabiting dinosaurs probably qualify here. Exactly when you pass the point where a once good idea becomes not worth the trouble is at first a matter of judgment. There certainly have been cases in which well-trained, smart people disagreed. But eventually a point is reached where there is such a preponderance of evidence that no rational, fair-minded person will think the idea plausible.

One way to assess whether you've reached that point is to look at uniqueness. During a scientific revolution, several proposals for unification are often on the table at any given time, threatening to take science in incompatible directions. This is normal, and in the midst of the revolution there does not need to be a rational reason to choose one over the others. At such times, even very smart people who choose between competing views too soon will often be wrong.

But one proposal for unification may end up explaining far more than the others, and it is usually the simplest. At this point, when a single proposal is vastly superior to others in terms of generation of new insights, agreement with experiment, explanatory power, and simplicity, it takes on an appearance of uniqueness. We say it has the ring of truth.

To see how this can happen, let us consider three unifications proposed by one person, the German astronomer Johannes Kepler (1571–1630). Kepler's lifelong obsession was the planets. Since he believed that the earth was a planet, he knew of six, Mercury out to Saturn. Their motions on the sky had been observed for thousands of years, so there was a lot of data. The most accurate came from Tycho Brahe, a Danish astronomer. Kepler eventually went to work for Tycho to get hold of his data (and after Tycho died he stole it, but that is another story).

Each planetary orbit has a radius. Each planet also has an orbital speed. In addition, the speeds are not uniform; the planets speed up and slow down as they move around the sun on their orbits. All of these numbers seem arbitrary. Kepler had been seeking his whole life for a principle that would unify the motions of the planets and, by doing so, explain the data of the planetary orbits.

Kepler's first try at a unification of the planets was in line with an ancient tradition that cosmological theory must employ only the simplest figures. One reason the Greeks had believed in circles moving on circles is that the circle is the simplest and hence, to them, the most beautiful of closed curves. Kepler searched for equally beautiful geometrical figures that might explain the sizes of the orbits of the planets. And he had a very elegant idea, illustrated in Fig. 1.

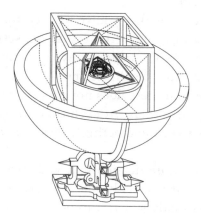

Fig. 1. Kepler's first theory of the solar system, based on the Platonic solids.

Let us take the orbit of Earth as given. There are, then, five numbers to explain: the ratio of the diameters of the orbits of the other five planets to the diameter of Earth's orbit. If they are to be explained, there must be some beautiful geometric construction that yields exactly five numbers. No more and no less. So is there a problem in geometry to which there are exactly five answers?

Yes. The cube is a perfect kind of solid, for each side is the same as every other side, and each edge is the same length as all the other edges. Such solids are called Platonic solids. How many are there? Exactly five: besides the cube, there is the tetrahedron, the octahedron, the dodecahedron, and the icosahedron.

It didn't take Kepler long to make an amazing discovery. Embed the orbit of Earth in a sphere. Fit a dodecahedron around the sphere. Put a sphere over that. The orbit of Mars fits on that sphere. Put the tetrahedron around that sphere, and another sphere around the tetrahedron. Jupiter fits on that sphere. Around Jupiter's orbit is the cube, with Saturn beyond. Inside Earth's orbit, Kepler placed the icosahedron, about which Venus orbited, and within Venus's orbit was the octahedron, for Mercury.

This unified theory explained the diameters of the orbits of the planets, something no theory had done before. It was mathematically beautiful. So why wasn't it believed? As compelling as it was, it didn't lead anywhere. No new phenomena were predicted on its basis. It didn't even lead to an understanding of the planets' orbital speeds. The idea was too static; it unified, but it didn't take science anywhere interesting.

Kepler thought about this for a long time. Since the diameters of the orbits were explained, he just needed to explain the speeds of the different planets. Finally he proposed that as the planets travel they "sing," and the frequencies of the notes are proportional to their speeds. The pitches sung by the different planets as they travel in their orbits make a harmony in six voices, which he called *the harmony of the spheres.*

This idea also has ancient roots, harking back to Pythagoras's discovery that the roots of musical harmony are in ratios of numbers. But it suffers from clear problems. It is not unique: There are many beautiful harmonizations of six voices. Even worse, there turned out to be more than six planets. And Galileo, a contemporary of Kepler, discovered four moons orbiting Jupiter. So there was yet another system of orbits in the sky. If Kepler's theories were right, they should apply to the newly discovered system. But they didn't.

Apart from those two proposals for the mathematical structure of the cosmos, Kepler made three discoveries that did lead to real progress in science. These were the three laws he is now famous for, proposed after years spent painstakingly analyzing the data he stole from Tycho. They have none of the beauty of his other proposals, but they do work. Moreover, one of them accomplishes something he could do no other way, which is to find a relationship between the speeds and the diameters of the orbits. Kepler's three laws not only agreed with the data on all six planets, they agreed with observations of Jupiter's moons.

Kepler discovered these laws because he took Copernicus's unification to its logical conclusion. Copernicus had said that the sun was at (or actually, near) the center of the universe, but in his theory the planets would move the same way whether the sun was there or not. Its only role was to light up the scene. The success of Copernicus's theory led Kepler to ask whether the sun's being near the center of each planet's orbit could really be a coincidence. He wondered whether the sun might instead play some role in causing the planetary orbits. Might the sun in some way exert a force on the planets, and might that force be the explanation for their motion?

To answer these questions, Kepler had to find a role for the precise position of the sun in each orbit. His first big breakthrough was to discover that the orbits were not circles, they were ellipses. And the sun had an exact role: It was exactly at the focus of the ellipse of each orbit. This was his first law. Shortly after this, he discovered his second law, which was that the speed of a planet in its orbit increased or decreased as it moved closer to or farther from the sun. He later discovered his third law, which governed how the speeds of the planets were related.

These laws point to some deep fact unifying the solar system, because the laws apply to all the planets. The payoff is that for the first time we had a theory that could make predictions. Suppose a new planet is discovered. Can we predict what its orbit will be? Before Kepler, no one could. But given Kepler's laws, all we need is two observations of its position and we can predict its orbit.

These discoveries paved the way for Newton. It was Newton's great insight to see that the force the sun exerted on the planets is the same as the force of gravity that holds us on Earth, and hence to unify physics in the heavens with physics on Earth.

Of course, the idea of a force emanating from the sun to the planets was absurd to most scientists at the time. They believed that space was empty; there was no medium that could convey such a force. Furthermore, there was no visible manifestation of it — no arm reaching out from the sun to each planet — and nothing invisible could be real.

There are good lessons here for would-be unifiers. One is that mathematical beauty can be misleading. Simple observations made from the data are often more important. Another lesson is that correct unifications have consequences for phenomena unsuspected at the time a unification is invented, as in the case of the application of Kepler's laws to Jupiter's moons. Correct unifications also raise questions that may seem absurd at the time but lead to further unifications, as in Kepler's postulation of a force from the sun to the planets.

Most important, we see that a real revolution often requires that several new proposals for unification come together to support one another. In the Newtonian revolution, there were several proposed unifications that triumphed at once: the unification of the earth with the planets, the unification of the sun with the stars, the unification of rest and uniform motion, and the unification of the gravitational force on Earth with the force by which the sun influences a planet's motions. Singly, none of these ideas could have survived; together, they trounced their rivals. The result was a revolution that transformed every aspect of our understanding of nature.

In the history of physics, there is one unification that serves more than any other as a model for what physicists have been trying to do in the last thirty years. This is the unification of electricity and magnetism, achieved by James Clerk Maxwell in the 1860s. Maxwell made use of a powerful idea called a field, which had been invented by the British physicist Michael Faraday in the 1840s to explain how a force could be conveyed through empty space from one body to another. The idea is that a field is a quantity, like a number, one of which lives at each point in space. As you move through space, the value of the field changes continuously. The value of the field at a single point also evolves in time. The theory gives us laws that tell us how the field changes as you move in space and through time. These laws tell us that the value of the field at a particular point is influenced by the value of the field at nearby points. The field at a point can also be influenced by a material body at the same point. Thus, a field can carry a force from one body to another. There is no need to believe in ghostly action at a distance.

One field Faraday studied was the electric field. This is not a number but a vector, which we may visualize as an arrow and which can vary its direction and length. Imagine such an arrow at each point of space. Imagine that the ends of the arrows at nearby points are attached to one another by rubber bands. If I pull on one, it pulls on the ones nearby. The arrows are also influenced by electric charges. The effect of the influence is that the arrows will arrange themselves so that they point to nearby negative charges and away from nearby positive charges.

Faraday also studied magnetism. He invented another field, an-

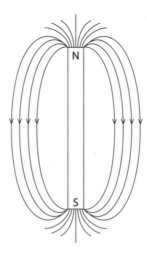

*Fig. 2. Lines of force trace the magnetic field
arising from a bar magnet.*

other collection of arrows, which he called the magnetic field; these
arrows like to point to poles of magnets (see Fig. 2).

Faraday wrote down simple laws to describe how the electric and
magnetized field arrows are influenced by nearby charges and magnetic poles and also by the arrows of nearby fields. He and others
tested the laws and found they gave predictions that agreed with experiment.

Among the discoveries of the time were phenomena that mixed
electric and magnetic effects. For example, a charge moving in a circle gave rise to magnetic fields. Maxwell realized that these discoveries pointed to a unification of electricity and magnetism. To fully
unify them, he had to change the equations. When he did so, simply
by adding one term, his unification became a unification with consequences.

The new equations allowed electric and magnetic fields to turn
into each other. These transmutations give rise to waves of shift-
ing patterns, in which first there is an electric field and then a magnetic field, and which move through space. Such moving patterns
could be made by, among other things, waving an electric charge
back and forth. The ensuing waves could carry energy from one
place to another.

The most amazing thing was that Maxwell could compute the speed of these waves from this theory, and he found that they were the same as the speed of light. Then it must have hit him. *The waves passing through the electric and magnetic fields are light.* Maxwell did not set out to make a theory of light, he set out to unify electricity and magnetism. But in doing so, he achieved something greater. This is an example of how a good unification will have unexpected consequences for both theory and experiment.

New predictions immediately followed. Maxwell realized that there should be electromagnetic waves at all frequencies, not just those of visible light, and this led to the discovery of radio, infrared light, ultraviolet light, and so on. This illustrates another historical lesson: When someone proposes the right new unification, the implications become obvious very quickly. Many of these phenomena were observed in the first years after Maxwell published his theory.

This raises a point that will become important when we discuss other proposals for unification. All unifications have consequences because they lead to phenomena that arise because the things that were unified can transform into one another. In the good cases, these new phenomena are soon observed — the inventors have every right to celebrate the unification. But we will see that in other cases the predicted phenomena are already in conflict with observation. In this unhappy event, the proponents have to either give up their theory or constrain it unnaturally so as to hide the consequences of the unification.

But even as it triumphed, Maxwell's unification of electricity and magnetism faced one formidable obstacle. In the mid-nineteenth century, most physicists believed that physics was unified because everything was made of matter (and had to be, in order to satisfy Newton's laws). For these "mechanists," the idea of a field just waving in space was hard to swallow. Maxwell's theory made no sense to them without some stuff whose bending and stretching would constitute the true reality behind the electric and magnetic fields. Something material must be quivering when a light wave travels from a flower to one's eye.

Faraday and Maxwell were themselves mechanists, and they devoted a lot of time and trouble to addressing this problem. They were not alone; young gentlemen made good careers at renowned institutions by inventing elaborate constructions of the microscopic gears, pulleys, and belts that they posited underlay Maxwell's equations. Prizes were given for those who could solve the convoluted equations that resulted.

There was one big and obvious manifestation of the problem, which is that light travels to us from the sun and stars, and outer space is empty of any matter. Were there any matter in space, it would retard the motion of the planets, which would thus have long since fallen into the sun. But how could electric and magnetic fields reside in a vacuum?

So the mechanists invented a new form of matter — the *aether* — and filled space with it. The aether had paradoxical properties: It had to be extremely dense and stiff, for light was to be essentially a sound wave through it. The huge ratio of the speed of light to that of sound had to be a consequence of the incredible density of the aether. At the same time, the aether had to offer absolutely no resistance to the passage of ordinary matter through it. This is harder to arrange than it looks. One can just say that the aether and ordinary matter don't interact with each other — that is, that they exert no forces on each other. But then why should ordinary matter detect light — or electric or magnetic fields — if these are just stresses in the aether? No wonder professorships were given to those who cleverly worked it all out.

Could there have been a more beautiful unification than the aether theory? Not only were light, electricity, and magnetism unified, their unification was unified with matter.

However, while the aether theory was being developed, the physicists' conception of matter was also changing. In the early nineteenth century, most physicists had thought of matter as continuous, but electrons were discovered late in the century, and the idea that matter is made of atoms was taken more seriously then — at least, by some physicists. But that raised another question: What were atoms and electrons in a world made of aether?

Picture field lines, like the lines of a magnetic field running from the north pole to the south pole of a magnet. The field lines can never end, unless they end on the pole of a magnet; this is one of Maxwell's laws. But they can make closed circles, and those circles can tie themselves up in knots. So perhaps atoms are knots in magnetic field lines.

But as every sailor knows, there are different ways to tie a knot. Maybe that's good, because there are different kinds of atoms. In 1867 Lord Kelvin proposed that the various atoms would correspond to different knots.

This may seem absurd, but recall that at the time we knew very little about atoms. We knew nothing about nuclei and had never heard of protons or neutrons. So this was not as crazy as it might seem.

At that time, we also knew very little about knots. No one knew how many ways there were to tie a knot or how to tell them apart. So, inspired by this idea, mathematicians began studying the problem of how to distinguish the various possible knots. This slowly turned into a whole field of mathematics called *knot theory*. It soon was proved that there are an infinite number of distinct ways to tie a knot, but it has taken a long time to learn how to tell them apart. Some progress was made in the 1980s, but there is still no known procedure for telling whether two complicated knots are the same or different.

Notice how a good idea of unification, even if it turns out to be wrong, can inspire new avenues of inquiry. We should keep in mind, though, that just because a unified theory is fruitful for mathematics does not mean that the physical theory is correct. Otherwise, the success of knot theory would require us to still believe that atoms are knots in a magnetic field.

There was a further problem: Maxwell's theory appeared to contradict the principle of relativity from Newtonian physics. It turned out that by doing various experiments, including measuring the speed of light, observers studying an electromagnetic field could tell whether they were moving or not.

Here is a conflict between two unifications, both central to Newton's physics: the unification of everything as matter obeying Newton's laws versus the unification of motion and rest. For many physicists, the answer was obvious: The idea of a material universe was more important than the perhaps accidental fact that it was hard to detect motion. But a few took the principle of relativity as more important. One of these was a young student studying in Zurich called Albert Einstein. He meditated on the puzzle for ten years, beginning at the age of 16, and finally, in 1905, realized that the resolution required a complete revision of our understanding of space and time.

Einstein solved the puzzle by playing the same great trick that Newton and Galileo had originally played to establish the relativity of motion. He realized that the distinction between electrical and magnetic effects depends on the motion of the observer. So Maxwell's unification was deeper than even Maxwell had suspected. Not only were the electric and magnetic fields different aspects of a single phenomenon, but different observers would draw the distinction differently; that is, one observer might explain a particular phenomenon in terms of electricity, while another observer, moving relative to the first, would explain the same phenomenon in terms of magnetism. But the two would agree about what was happening. And so Einstein's special theory of relativity was born, as a joining of Galileo's unification of rest and motion with Maxwell's unification of electricity and magnetism.

Much follows from this. One consequence is that light must have a universal speed, independent of the motion of the observer. Another is that there must be a unification of space and time. Previously, there had been a clear distinction: Time was universal, and everyone would agree on what it meant for two things to happen simultaneously. Einstein showed that observers in motion with respect to each other would disagree about whether two events at different places were happening at the same time or not. This unification was implicit in his 1905 paper titled "On the Electrodynamics of Moving Bodies," and it was stated explicitly in 1907 by one of his teachers, Hermann Minkowski.

So here we have again a story of two competing attempts at unification. The mechanists had a beautiful idea that unified physics: Everything is matter. Einstein believed in another kind of unification, the unification of motion and rest. To support this, he had to invent a still deeper unification — of space and time. In each case, something previously thought to be absolutely distinct becomes distinct only relative to the motion of the observer.

In the end, the conflict between the two proposals for unification was settled by experiment. If you believed the mechanists, you believed that an observer could measure his speed through the aether. If you believed Einstein, you knew that he couldn't, as all observers are equivalent.

Several attempts had been made to detect Earth's motion through the aether before 1905, when Einstein proposed special relativity, and they had failed.[1] Proponents of the aether theory had just adjusted their predictions so as to make it harder and harder to detect Earth's motion. This was easy to do, because when they did calculations they used Maxwell's theory, which, when correctly interpreted, agreed with Einstein's expectations that motion was not detectable. That is, the mechanists already had the right equations, they just had the wrong interpretations.

As for Einstein himself, it's not clear how much he knew about the early experiments, but they wouldn't have mattered to him, as he was already convinced that the motion of the earth was not detectable. Einstein was in fact only getting started. As we shall see in the next chapter, his unification of space and time was about to deepen considerably. By the time most physicists had caught up with him and accepted the special theory of relativity, Einstein was already moving far beyond it.

3

The World
As Geometry

THE EARLY DECADES of the twentieth century saw several attempts at unification. A few succeeded, the rest failed. By briefly telling their stories, we can draw lessons that will help us understand the crisis facing the current attempts at unification.

From Newton to Einstein, a single idea dominated: *The world is made of nothing but matter.* Even electricity and magnetism were aspects of matter — just stresses in the aether. But this beautiful picture was crushed when special relativity triumphed, for if the whole notion of being at rest or in motion is meaningless, the aether must be a fiction.

The quest for unification had to go somewhere, and there was really only one place to go. This was to reverse the aether theory: If fields are not made from matter, perhaps *fields* are the fundamental stuff. *Matter must then be made from fields.* There were already models of electrons and atoms as stresses in the fields, so this was not such a big step.

But even as this idea gained adherents, there were still mysteries. For example, there are two different kinds of fields, the gravitational field and the electromagnetic field. Why two fields and not a single field? Is this the end of the story? The yearning for unification compelled physicists to ask whether gravity and electromagnetism were aspects of a single phenomenon. Thus was born the search for what we now call a *unified-field theory*.

Since Einstein had just incorporated electromagnetism into his special theory of relativity, the most logical way to proceed was to modify Newton's theory of gravity so as to make it consistent with relativity theory. This turned out to be easy to do. Not only that, this modification led to a wonderful discovery that would become the core of unified theories to this day. In 1914, a Finnish physicist named Gunnar Nordström found that all you had to do to unify gravity with electromagnetism was increase the dimensions of space by one. He wrote the equations that describe electromagnetism in a world with four dimensions of space (and one of time), and out popped gravity. Just by the extra dimension of space, you got a unification of gravity with electromagnetism that was also perfectly consistent with Einstein's special theory of relativity.

But if this is true, shouldn't we be able to look out in this new dimension, as we look out in the three dimensions of space? If not, isn't this theory then obviously wrong? To avoid this troublesome issue, we can make the new dimension a circle, so that when we look out, we in effect travel around it and come back to the same place.[1] Then we can make the diameter of the circle very small, so that it is hard to see that the extra dimension is there at all. To understand how shrinking something can make it impossible to see, recall that light is made up of waves and each light wave has a wavelength, which is the distance between peaks. The wavelength of a light wave limits how small a thing you can see, for you cannot resolve an object smaller than the wavelength of the light you use to see it. Hence, one cannot detect the existence of an extra dimension smaller than the wavelength of light one can perceive.

One might think that Einstein, of all people, would have embraced this new theory. But by that time (1914), he was already traveling down a different road. Unlike his contemporaries, Einstein had taken a route to the unification of gravity with relativity that brought him back to the very foundation of the principle of relativity: the unification of motion and rest discovered by Galileo several centuries earlier. That unification involved only uniform motion — that is, motion in a straight line at a constant speed. Beginning around 1907, Einstein started to ask himself about other types of motion, such as accelerated motion. This is motion whose speed or direction changes. Shouldn't the distinction between accelerating and nonaccelerating motion somehow be erased?

At first this seems a misstep, for while we can't feel the effects of uniform motion, we certainly do feel the effects of acceleration. When an airplane takes off, we feel pushed back into our seats. When an elevator begins to rise, we feel the acceleration in the form of additional pressure pushing us into the floor.

It was at this point that Einstein had his most extraordinary insight. He realized that *the effects of acceleration were indistinguishable from the effects of gravity*. Think of a woman standing in an elevator waiting for it to move. She already feels a force pulling her to the floor. What happens when the elevator starts to ascend is not different in kind, only in degree: She feels the same force increase. Suppose that the elevator stays still but the strength of gravity increases momentarily? Einstein realized that she would feel exactly the same as if the elevator had accelerated upward.

There is a converse to this. Suppose that the cable holding the elevator is cut and the car, with its occupants, begins to fall. In free fall, the occupants of the elevator will feel weightless. They will feel exactly as astronauts do in orbit. That is to say, the acceleration of the falling elevator can exactly cancel the effects of gravity.

Einstein recalled realizing that a person falling from the roof of a building would feel no effects of gravity as he fell. He called this "the most fortunate thought of my life," and he made it into a principle, which he called *the principle of equivalence*. It says that the effects of acceleration are indistinguishable from the effects of gravity.[2]

So Einstein succeeded in unifying all kinds of motion. Uniform motion is indistinguishable from rest. And acceleration is no different from being at rest but with a gravitational field turned on.

The unification of acceleration with gravity was a unification with great consequences. Even before the conceptual implications were worked out, there were huge implications for experiment. Some predictions could even be derived with high school algebra — for example, that clocks would slow down in a gravitational field, which was eventually confirmed. Another prediction — first made by Einstein, in 1911 — was that light is bent when it passes through a gravitational field.

Notice here that, as in the successful unifications discussed earlier, more than one unification is happening at once. Two different kinds of motion are being unified; there is no longer a need to distinguish uniform from accelerated motion. And the effects of acceleration are being unified with the effects of gravity.

Even if Einstein could reason from the equivalence principle to a few predictions, the new principle was not a complete theory. The formulation of a complete theory was the greatest challenge of his life and took nearly a decade to accomplish. To see why, let us try to understand what it means to say that gravity bends light rays. Before this particular insight of Einstein's, there had always been two different kinds of things in the world: the things that live in space and space itself.

We are not accustomed to thinking of space as an entity with properties of its own, but it certainly is. Space has three dimensions, and it also has a particular geometry, which we learn in school. Called Euclidean geometry — after Euclid, who worked out its postulates and axioms more than two thousand years ago — it is the study of the properties of space itself. The theorems of Euclidean geometry tell us what happens to triangles, circles, and lines drawn in space. But they hold for all objects, material or imagined.

A consequence of Maxwell's theory of electromagnetism is that light rays move in straight lines. Thus it makes sense to use light rays when tracing the geometry of space. But if we adopt this idea, we see immediately that Einstein's theory has great implications. For light rays are bent by gravitational fields, which, in turn, respond to the presence of matter. The only conclusion to draw is that the presence of matter affects the geometry of space.

In Euclidean geometry, if two straight lines are initially parallel, they can never meet. But two light rays that are initially parallel can meet in the real world, because if they pass on each side of a star, they will be bent toward each other. So Euclidean geometry is not true in the real world. Moreover, the geometry is constantly changing, because matter is constantly moving. The geometry of space is not like a flat, infinite plane. It is like the surface of the ocean — incredibly dynamic, with great waves and small ripples in it.

Thus, the geometry of space was revealed to be just another field. Indeed, the geometry of space is almost the same as the gravitational field. To explain why, we have to recall the partial unification of space and time that Einstein achieved in special relativity. In this unification, space and time together make up a four-dimensional entity called spacetime. This has a geometry analogous to Euclidean geometry, in the following precise way.

Consider a straight line in space. Two particles can travel along it, but one travels at a uniform speed, while the other is constantly accelerating. As far as space is concerned, the two particles travel on the same path. *But they travel on different paths in spacetime.* The particle with a constant speed travels on a straight line, not only in space but also in spacetime. The accelerating particle travels on a curved path in spacetime (see Fig. 3).

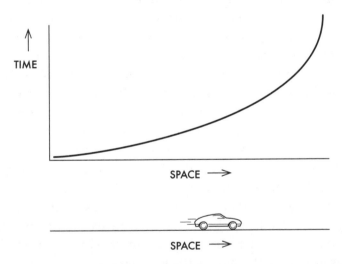

Fig. 3. A car decelerating along a straight line in space travels on a curved path in spacetime.

Hence, just as the geometry of space can distinguish a straight line from a curved path, the geometry of spacetime can distinguish a particle moving at a constant speed from one that is accelerating.

But Einstein's equivalence principle tells us that the effects of gravity cannot be distinguished, over small distances, from the effects of acceleration.[3] Hence, by telling which trajectories are accelerated and which are not, the geometry of spacetime describes the effects of gravity. The geometry of spacetime is therefore the gravitational field.

Thus, the double unification given by the equivalence principle becomes a triple unification: All motions are equivalent once the effects of gravity are taken into account, gravity is indistinguishable from acceleration, and the gravitational field is unified with the geometry of space and time. When worked out in detail, this became Einstein's *general theory of relativity*, which he published in full form in 1915.

Not bad for a guy who couldn't initially get an academic job.

Thus, by 1916 there were two very different proposals for the future of physics, both based on a deep idea about unifying gravity with the rest of physics. There was Nordström's elegant unification of gravity with electromagnetism by the simple postulation of an extra, hidden dimension of space. And there was Einstein's general theory of relativity. Both seemed consistent theories and each did something unexpectedly elegant.

They could not both be true, so a choice had to be made. Fortunately, the two theories made different predictions for a doable experiment. Einstein's general theory of relativity predicted that gravity must bend light rays — and by precisely how much. In Nordström's theory, there was no such effect: Light always goes in straight lines, period.

In 1919, the great British astrophysicist Arthur Eddington led an expedition off the west coast of Africa to conduct an experiment, which ended by confirming that the gravitational field of the sun indeed bends light. This effect was observed during a total solar eclipse, which made it possible to see, near the rim of the occluded sun, the light from stars that were actually directly behind the sun. Had the sun's gravity not bent their light, these stars would not have been visible. But they were. So the choice between the two profoundly different directions for unification was made in the only way it could have been — by experiment.

This is an important example, because it shows the limits of what can be accomplished by thought alone. Some physicists have argued that general relativity is a case in which pure thought sufficed to show the way forward. But the real story is the opposite. Without experiment, most theorists would probably have chosen Nordström's unification; it is simpler and brought with it the powerful new idea of unification through extra dimensions.

Einstein's unification of the gravitational field with the geometry of spacetime signaled a profound transformation in how we conceive of nature. Before Einstein, space and time were thought to have properties that were fixed for all eternity: The geometry of space is, was, and always would be as Euclid described. Time marched on independently of anything else. Things moved in space and evolved over time, but space and time themselves never altered.

For Newton, space and time constituted an absolute background. They provided a fixed stage on which a grand drama is played out. The geometry of space and time was needed to give meaning to the things that change, like the positions and motions of particles. But they themselves never changed. We have a name for theories of physics that rely on such an absolute, fixed framework: We call them *background-dependent theories*.

Einstein's general theory of relativity is completely different. There is no fixed background. The geometry of space and time changes and evolves, as does everything else in nature. Different geometries of spacetime describe the histories of different universes. We no longer have fields moving in a fixed-background geometry. We have a bunch of fields all interacting with one another, all dynamical, all influencing one another, one of which is the geometry of spacetime. We call such a theory a *background-independent theory*.

Make note of the distinction between background-dependent and background-independent theories. The story that unfolds over the course of this book turns on the difference between them.

Einstein's general theory of relativity satisfied all the tests we laid out in the last chapter for a successful unification. There were profound conceptual consequences, which were implied by the unifications involved. These quickly led to predictions of new phenomena, such as the expanding universe, the Big Bang, gravitational waves, and black holes, and there is good evidence for all of them. Our whole notion of cosmology was turned on its head. Proposals that once seemed radical, like the bending of light by matter, are now used as tools to trace the distribution of matter in the universe.

And every time the predictions of the theory are tested in detail, they are beautifully borne out.[4]

But general relativity was just the start. Even before Einstein had published the final version of the theory, he and others were formulating new kinds of unified theories. They had in common a simple idea: If the gravitational force could be understood as a manifestation of the geometry of space, why couldn't this also be true of electromagnetism? In 1915, Einstein had written to David Hilbert, perhaps the greatest mathematician then living, "I have often tortured my mind in order to bridge the gap between gravitation and electromagnetism."[5]

But it took until 1918 for a really good idea about this particular unification to emerge. This theory, invented by the mathematician Herman Weyl, contained a beautiful mathematical idea that was to become the core of the standard model of particle physics. Yet it failed, because in Weyl's original version it had big consequences that disagreed with experiment. One was that the length of an object would depend on the path it took. If you took two meter sticks, separated them, and then brought them back together and compared them, they would in general be different in length. This is more radical then special relativity, which holds that meter sticks can indeed appear to have different lengths, but only when they are moving relative to each other, not when they are compared at rest. It also, of course, disagrees with our experience of nature.

Einstein didn't believe Weyl's theory, but he admired it, writing to Weyl, "Apart from the [lack of] agreement with reality it is in any case a superb intellectual performance."[6] Weyl's reply shows the power of mathematical beauty: "Your rejection of the theory for me is weighty, . . . But my own brain still keeps believing in it."[7]

The tension between those caught in the allure of a beautiful theory they invented and more sober minds insisting on a connection to reality is a story we will see repeated in later attempts at unification. There is no easy resolution in these cases, because a theory can be fantastically beautiful, fruitful for the development of science, and yet at the same time completely wrong.

But even though Weyl's first try at unification failed, he had invented the modern concept of unification that would lead eventually to string theory. He was the first, but far from the last, to proclaim, "I am bold enough to believe that the whole of physical phenomena may be derived from one single universal world-law of the greatest mathematical simplicity."[8]

A year after Weyl's theory, a German physicist named Theodor Kaluza found a different way to unify gravity and electromagnetism, by reviving Nordström's idea of the hidden dimension. But he did it with a twist. Nordström had found gravity by applying Maxwell's theory of electromagnetism to a five-dimensional world (of which four dimensions are spatial and one is time). Kaluza did this in reverse: He applied Einstein's general theory of relativity to a five-dimensional world and found electromagnetism.

You can visualize this new space by attaching a little circle to each point of ordinary three-dimensional space (see Fig. 4). This new geometry can curve in new ways, because the little circles can be attached differently at different points. There is then something new to measure at each point of the original three-dimensional space. This information turns out to look just like the electric and magnetic fields.

Another wonderful spin-off is that it turns out that the charge of the electron is related to the radius of the little circle. This should not be surprising: If the electric field is just a manifestation of geometry, the electric charge should be, too.

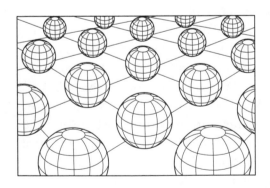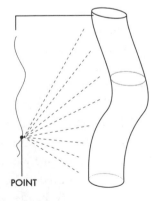

POINT

Fig. 4. Curled-up extra dimensions used in Kaluza-Klein theory. Left: A sphere is placed over every point of ordinary three-dimensional space, making a five-dimensional space. Right: A small circle is placed over a one-dimensional space. From far away, the space looks one-dimensional, but examined closely it is seen to be two-dimensional.

And not only this. General relativity describes the dynamics of the geometry of spacetime in terms of certain equations, called the Einstein equations. I don't need to write them down to describe the key fact: Those same equations can be applied to the five-dimensional world we just described. As long as we impose one simple condition, they turn out to be the right equations to describe the electric and magnetic fields and gravity, unified together. Thus, if this theory is right, the electromagnetic field is just another name for the geometry of the fifth dimension.

Kaluza's idea was rediscovered and further developed in the 1920s by the Swedish physicist Oskar Klein. Their theory was beautiful and compelling indeed. Gravity and electromagnetism are unified in one blow, and Maxwell's equations are explained as coming out of Einstein's equations, all by the simple act of adding a single dimension to space.

This time Einstein was enthralled. In April 1919 he wrote to Kaluza, "The idea of achieving [a unified theory] by a five-dimensional cylinder had never occurred to me. . . . At first glance I like your idea enormously."[9] In a letter some years later to Dutch physicist Hendrik Lorentz, he exulted, "It appears that the union of gravitation and Maxwell's theory is achieved in a completely satisfactory way by the five-dimensional theory."[10] George Uhlenbeck, a prominent physicist, remembered first hearing Klein's idea in 1926: "I felt a kind of ecstasy! Now one understands the world."[11]

Unfortunately, Einstein and the other enthusiasts were wrong. As with Nordström's theory, the idea of unification by adding a hidden dimension failed. It is important to understand why.

I said earlier that for a proposed unification to succeed, it has to win its place by making new predictions that are confirmed by experiment. Successful unifications also generate a plethora of new insights that lead to further discoveries. Compelling as it was to some, neither of these things happened in the case of Kaluza-Klein theory. The reason is simple: The theory imposed an extra condition, referred to earlier, which is that the extra dimension is curled up into a circle whose radius is too small to see. Not only that: To get electromagnetism out of the theory, the radius of the circle must be frozen, changing in neither space nor time.

This is the Achilles' heel of the whole enterprise and led directly to its failure. The reason is that freezing the radius of the extra dimension undermines the very essence of Einstein's theory of general relativity, which is that geometry is dynamical. If we add another dimension to spacetime as described by general relativity, the geometry of that extra dimension should also be dynamical. And indeed, it would be, were the radius of the little circle allowed to move freely. The theory of Kaluza and Klein would then have infinitely many solutions in which the radius of the circle varies over space and changes in time. This would have wonderful implications, because it would lead to processes in which gravitational and electrical effects convert into each other. It would also lead to processes in which electrical charges vary over time.

But if the Kaluza-Klein theory is a true unification, the fifth dimension cannot be treated differently from the others: The little circle must be allowed to change. The resulting processes are hence the necessary consequences of unifying electricity and geometry. If they were ever observed, they would confirm directly that geometry, gravity, electricity, and magnetism are all aspects of one phenomenon. Unfortunately, such effects have never been observed.

This is not one of those cases in which theorists can quickly celebrate the consequences of the unification; instead, they must hide it, by insisting on studying only an infinitesimal fraction of solutions where the radius of the fifth dimension is frozen in space and time.

It gets worse, because such solutions, it turns out, are unstable. Tickle the geometry just a bit, and the small circle collapses quickly to a singularity marking the end of time. Tickle it a different way and the circle grows, so that soon the extra dimension becomes visible, discrediting the theory entirely. As a result, the theory's predictions must be hidden to cover the fact that it gets so much wrong.

At this point, even Einstein lost his enthusiasm. He wrote to his friend Paul Ehrenfest, "It is anomalous to replace the four-dimensional continuum by a five-dimensional one and then subsequently to tie up artificially one of those five dimensions in order to account for the fact that it does not manifest itself."[12]

As if this were not enough, physicists had other reasons to reject the theory. By the 1930s, people knew that there were more forces in the world than gravity and electromagnetism. They knew about the strong and the weak nuclear forces, so it made no sense to leave these out of the unification. But no one knew how to include them in these unified theories. Still, for a while the search for a unified-field theory continued, led by Einstein. Some of the great mathematicians and physicists of the time contributed to this effort, including Wolfgang Pauli, Erwin Schrödinger, and Weyl. They found other ways to modify the geometry of spacetime so as to unify gravity with electromagnetism. These relied on deep mathematical insights, but they, too, led nowhere; they either made no new predictions or they predicted phenomena that were not seen. By the 1940s, Einstein and the few others who still pursued a unified-field theory were mostly laughed at.

My first job after getting my PhD was in 1979 at the Institute for Advanced Study, in Princeton. One of my main reasons for taking it was the hope of making contact with some living legacy of Einstein, who had died twenty-four years earlier. In this I was disappointed. There was no trace of his time there, apart from a bust of him in the library. No student or follower of Einstein could be found. Only a few people who had known him, like the theoretical physicist Freeman Dyson, were still there.

My first week there, Dyson, very much the gentleman, came by and invited me to lunch. After inquiring about my work, he asked if there was anything he could do to make me more at home in Princeton. I had but one request. "Could you tell me what Einstein was really like?" I asked. Dyson replied, "I'm very sorry, but that's one thing I can't help you with." Surprised, I insisted, "But you came here in 1947 and you were a colleague of his until he died in 1955."

Dyson explained that he too had come to the institute hoping to get to know Einstein. So he went to Einstein's secretary, Helen Dukas, to make an appointment. The day before the appointment, he began to worry about not having anything specific to discuss with the great man, so he got from Ms. Dukas copies of Einstein's recent scientific papers. They were all about Einstein's efforts to construct a unified-field theory. Reading them that evening, Dyson decided they were junk.

The next morning, he realized that although he couldn't face Einstein and tell him his work was junk, he couldn't *not* tell him either. So he skipped the appointment and, he told me, spent the ensuing eight years before Einstein's death avoiding him.

I could only say the obvious: "Don't you think Einstein could have defended himself and explained his motivation to you?"

"Certainly," Dyson replied. "But I was much older before that thought occurred to me."

One problem that Einstein and the other few unified-field theorists faced (besides the derision of the particle physicists) was that this kind of unification turned out to be too easy. Rather than being hard to find, unified-field theories were a dime a dozen. There were many different ways to achieve them and no reason to choose one over another. In decades of work, there was only one real advance: The problem of incorporating the two nuclear forces was solved. It turned out that all that was required was to add still more extra dimensions. The fields needed to describe the weak and strong nuclear forces pop out when several more new dimensions are added to general relativity. The story is much the same as Kaluza's attempt with electromagnetism: One has to freeze the geometry of the extra dimensions, making sure that their geometry never changes in time or varies in space, and one must make them too small to see. When all this is done correctly, the necessary equations (known as Yang-Mills equations) result from applying the equations of general relativity to the higher dimensions.

The fact that the Yang-Mills equations were hidden in higher-dimensional extensions of general relativity was not discovered until the 1950s, but their significance was not grasped until the 1970s, when we finally understood that these equations described the weak and strong nuclear forces. When people did finally make that connection, there were a few attempts at reviving the Kaluza-Klein idea, but they didn't go very far. By then, we had learned that nature lacked a certain symmetry — that of parity between left and right. Specifically, all neutrinos are what is referred to as left-handed (that is, the direction of their spin is always opposite to that of their linear momentum). This means that if you look at the world in a mirror, you will see a false world — one in which neutrinos are right-handed. So the world seen in a mirror is not a possible world. But this asymmetry turned out to be hard to explain in a world described by Kaluza-Klein theory.

Beyond that, the higher-dimensional theories continued to make no new predictions. The conditions we had to impose on the extra dimensions in order to get the physics we wanted were the theory's seeds of destruction. Indeed, the more dimensions you include, the higher the price you pay for freezing their geometry. The more dimensions, the more degrees of freedom — and the more freedom is accorded to the geometry of the extra dimensions to wander away from the rigid geometry needed to reproduce the forces known in our three-dimensional world. The problem of instability gets worse and worse.

Moreover, whenever there is more than one hidden dimension, there are many different ways to curl them up. Rather than there being just one — a circle — there are an infinite number of ways the higher dimensions can be curled up, so there are an infinite number of possible versions of the theory. How is nature to choose among them?

Over and over again in the early attempts at unifying physics through extra dimensions, we encounter the same story. There are a few solutions that lead to the world we observe, but these are unstable islands in a vast landscape of possible solutions, the rest of which are very unlike our world. And once conditions are imposed to eliminate those, there is no smoking gun — no consequence of the unification that has not yet been seen but might be, if experimentalists were to look for it. So there is nothing to celebrate and a lot to hide.

But there was an even more fundamental problem, which had to do with the relationship of the unified theories to the quantum theory. The early attempts at unified-field theories took place before quantum mechanics was completely formulated, in 1926. Indeed, a few of the quantum theory's proponents had interesting speculations about the relationship between the extra dimensions and quantum theory. But after 1930 or so, there was a split. Most physicists ignored the problem of unification and concentrated instead on applying quantum theory to a vast array of phenomena, from the properties of materials to the processes by which stars produce energy. At the same time, those few who persisted in working on the unified theories increasingly ignored quantum theory. These people (Einstein included) worked as if Planck, Bohr, Heisenberg, and

Schrödinger had never existed. They were living after the quantum-mechanical revolution but pretending to work in an intellectual universe in which that revolution had never occurred. They seemed to their contemporaries like the quaint communities of Russian émigré aristocrats who, in the 1920s and 1930s, carried on their elaborate social rituals in Paris and New York as if they were back in tsarist Saint Petersburg.

Of course, Einstein was not just some washed-up émigré intellectual from a lost world (even if he *was* an émigré intellectual from a lost world). He knew full well that he was ignoring quantum theory, but he had a reason: He didn't believe in it. Even though he had sparked the quantum revolution with his realization that the photon was real, he rejected the outcome. He was hoping to discover a deeper theory of quantum phenomena that would be acceptable to him. That is exactly where he hoped his unified-field theory would lead him.

But it didn't. Einstein's dream of an end run around quantum theory failed, and it more or less died with him. By that time, few respected him and fewer followed him. The physicists at the time thought they had better things to do than play with fanciful ideas about unification. They were hard at work cataloging the many new particles that were being discovered and honing the theories of the two newly discovered fundamental forces. That someone might speculate that the world had more than three dimensions of space, curled up too small too see, seemed to them as crazy and unproductive as studying UFOs. There were no implications for experiment, no new predictions, so, in a period when theory developed hand in hand with experiment, no reason to pay attention.

But suppose for a moment that despite all the obstacles, we still wanted to take ideas of a unified field seriously. Could these theories be formulated in the language of quantum theory? The answer was resoundingly no. No one knew at that time how to make even general relativity consistent with quantum theory. All early attempts to do this failed. When you added more dimensions, or more twists to the geometry, things always got worse, not better. The larger the number of dimensions, the faster the equations spun out of control, spiraling into infinite quantities and inconsistencies.

So, while the idea of unification by invoking higher dimensions was very attractive, it was abandoned, and for good reason. It made no testable predictions. Even if such a theory produced special solutions that did describe our world, there were, as noted, many more that did not. And the few that did were unstable and could easily evolve into singularities, or into worlds quite unlike our own. And finally, they could not be made consistent with quantum theory. Keep these reasons in mind — because, again, the success or failure of the newer proposals for unification, such as string theory, depend on whether they can solve these very problems.

By the time I began my study of physics in the early 1970s, the idea of unifying gravity with the other forces was as dead as the idea of continuous matter. It was a lesson in the foolishness of once great thinkers. Ernst Mach didn't believe in atoms, James Clerk Maxwell believed in the aether, and Albert Einstein searched for a unified-field theory. Life is tough.

4

Unification Becomes
a Science

After the idea of unifying all four fundamental forces by inventing new dimensions failed, most theoretical physicists gave up on the idea of relating gravity to the other forces, a decision that made sense because gravity is so much weaker than the other three. Their attention was drawn instead to the zoo of elementary particles that the experimentalists were discovering in their particle accelerators. They searched the data for new principles that could at least unify all the different kinds of particles.

Ignoring gravity meant taking a step backward, to the understanding of space and time before Einstein's general theory of relativity. This was a dangerous thing to do in the long run, as it meant working with ideas that had already been superseded. But there was also an advantage, in that this approach led to a great simplification of the problem. The chief lesson of general relativity was that there is no fixed-background geometry for space and time; ignoring this meant that you could simply choose the background. This sent us back toward a Newtonian point of view, in which particles and fields inhabit a fixed background of space and time — a background whose properties are fixed eternally. Thus, the theories that developed from ignoring gravity are background-dependent.

However, it was not necessary to go all the way back to Newton. One could work within the description of space and time given by

Einstein's 1905 special theory of relativity. According to it, the geometry of space is that given by Euclid, which many of us study in junior high school; however, space is mixed with time, in order to accommodate Einstein's two postulates, the relativity of observers and the constancy of the speed of light. The theory cannot accommodate gravity, but it's the right setting for Maxwell's theory of the electric and magnetic fields.

Once quantum mechanics was fully formulated, the quantum theorists turned their attention to unifying electromagnetism with quantum theory. As the basic phenomena of electromagnetism are fields, the unification that would eventually result is called a *quantum field theory*. And because Einstein's special theory of relativity is the right setting for electromagnetism, these theories can also be seen as unifications of quantum theory with special relativity.

This was a much more challenging problem than applying quantum theory to particles, because a field has a value at every point of space. If we assume that space is continuous — which is what special relativity asserts — then there are a continuous infinity of variables. In quantum theory, each variable is subject to an uncertainty principle. One implication is that the more precisely you try to measure a variable, the more it fluctuates uncontrollably. An infinite number of variables fluctuating uncontrollably can easily get out of hand. When you ask the theory questions, you have to be very careful not to get infinite and inconsistent answers.

The quantum theorists already knew that for each electromagnetic wave there is a quantum particle, the *photon*. It took only a few years to work this out in detail, but the result was just a theory of photons moving freely; the next step would be to incorporate charged particles, such as electrons or protons, and describe how they interact with photons. The goal was a fully consistent theory of quantum electrodynamics, or QED. This was much more challenging. QED was first solved by the Japanese physicist Sin-Itiro Tomonaga during World War II, but the news did not reach the rest of the world until 1948 or so. By then, QED had been constructed twice more, independently, by the young Americans Richard Feynman and Julian Schwinger.

Once QED was understood, the task was to extend quantum field theory to the strong and weak nuclear forces. This would take another quarter century, and the key would be the discovery of two new principles: The first defined what electromagnetism and these nuclear interactions have in common. It is called the *gauge principle,* and as I will describe, it leads to a unification of all three forces. The second principle explains why, although unified, the three forces are so different. It is called *spontaneous symmetry breaking.* These two principles together form the cornerstone of the standard model of particle physics. Their precise application had to await the discovery that particles like the proton and neutron are not elementary after all; instead, they are made of quarks.

The proton and neutron each have three quarks, while other particles, called mesons, have two (more properly, a quark and an antiquark). This discovery was made in the early 1960s, independently, by Murray Gell-Mann at Caltech and George Zweig at CERN (the European Organization for Nuclear Research) in Geneva. Shortly afterward, James Bjorken, of the Stanford Linear Accelerator Center (SLAC), and Richard Feynman at Caltech proposed experiments that, when they were later carried out at SLAC, confirmed that the proton and neutron are indeed each made up of three quarks.

The discovery of the quarks was an essential step toward unification because the interaction of protons, neutrons, and other particles was exceedingly complicated. But there was a hope that the force between the quarks might itself be simple and that the observed complexities arise because protons and neutrons are composite objects. This kind of notion had proved true before: Whereas the forces between molecules are complicated, the forces between the atoms that make them up can be easily understood in terms of electromagnetism. The idea was for theorists to give up trying to understand the force between protons and neutrons in fundamental terms and ask instead how that force affected the quarks. This is reductionism at work — the old trick that the laws governing the parts are often simpler than those governing the whole — and it eventually paid off in the discovery of the deep commonality that connects the two nuclear forces, strong and weak, with electromagnetism. All three are consequences of the simple but powerful gauge principle.

The gauge principle is best understood in terms of something physicists refer to as a *symmetry.* Put simply, a symmetry is an operation that doesn't change how something behaves relative to the outside world. For example, if you rotate a ball, you don't change it; it's still a sphere. So when physicists talk of a symmetry, they can be referring to an operation in space, like rotation, that doesn't change the result of an experiment. But they can also be talking about any kind of change we make to an experiment that doesn't alter the outcome. For example, suppose you take two groups of cats — say, east-side cats and west-side cats — and you test their abilities in jumping. If there is no difference in the average jump a cat can make, then we say that cat-jumping is symmetric under the operation of trading all your east-side cats for west-side cats.

Here is another example, simplified and idealized to make the point. Consider an experiment in which a beam of protons is accelerated and then aimed at a target consisting of certain kinds of nuclei. You, as the experimenter, observe the pattern the protons make as they scatter off the nuclei in the target. Now, without changing the energy or the target, you substitute neutrons for protons. In certain cases, the pattern of scattering hardly changes. The experiment is said to reveal that the forces involved act the same on protons and neutrons. In other words, the act of replacing protons with neutrons is a *symmetry* of the forces between them and the nuclei in the target.

Knowing the symmetries is a good thing, because they tell you something about the forces involved. In the first example, we learn that the force of gravity on cats doesn't depend on where they come from; in the second, that certain nuclear forces can't tell the difference between a proton and a neutron. Sometimes all we get from a symmetry is such partial information about the forces. But there are special situations in which the symmetries completely determine the forces. This turns out to be the case for a class of forces called *gauge forces.* I won't bother you with exactly how this works, as we won't need it.[1] But the fact that all the properties of a force can be determined by knowing the symmetries is one of the most important discoveries of twentieth-century physics. This idea is what is meant by the gauge principle.[2]

There are two things we do need to know about the gauge principle. One is that the forces it leads to are conveyed by particles called *gauge bosons.* The other thing we need to know is that the electromagnetic, strong, and weak forces each turned out to be forces of this kind. The gauge boson that corresponds to the electromagnetic force is called the photon. Those that correspond to the strong force holding the quarks together are called *gluons*. Those that correspond to the weak force have a less interesting name — they are called, simply, *weak bosons*.

The gauge principle is the "beautiful mathematical idea," noted in chapter 3, that was discovered by Herman Weyl in his failed 1918 attempt at the unification of gravity and electromagnetism. Weyl is one of the deepest mathematicians ever to ponder the equations of physics, and it was he who understood that the structure of Maxwell's theory was entirely explained by a gauge force. In the 1950s, some people asked whether other field theories could be constructed using the gauge principle. It turned out that this could be done by basing them on symmetries involving the various kinds of elementary particles. These theories are now called Yang-Mills theories, after the names of two of their inventors.[3] At first no one knew what to do with these new theories. The new forces they described would have an infinite range, like electromagnetism. Physicists knew that the two nuclear forces each had a short range, so it did not seem that they could be described by a gauge theory.

What makes theoretical physics as much an art as a science is that the best theorists have a sixth sense about what results can be ignored. Thus, in the early 1960s, Sheldon Glashow, then a postdoc at the Niels Bohr Institute, suggested that the weak force was indeed described by a gauge theory. He simply posited that some unknown mechanism limited the range of the weak force. If this range problem could be solved, the weak force could then be unified with electromagnetism. But the overall problem would still have to be faced: How could you unify forces that manifest themselves as differently as electromagnetism and the strong and weak nuclear forces do?

This is an example of a general problem that plagues nearly every attempt at unification. The phenomena you hope to unify are different — otherwise there would be nothing surprising about their unification. So even if you discover some hidden unity, you still have to understand why and how it is that they appear to be different.

As we saw earlier, Einstein had a wonderful way of solving this problem for special and general relativity. He realized that the apparent differences between the phenomena were not intrinsic to the phenomena but were due entirely to the necessity of describing the phenomena from the viewpoint of an observer. Electricity and magnetism, motion and rest, gravity and acceleration were all unified by Einstein in this way. The differences that observers perceive between them are therefore contingent, because they reflect only the viewpoint of the observers.

In the 1960s, a different solution to this general problem was proposed: The differences between unified phenomena were contingent, but not because of the viewpoint of particular observers. Instead, physicists made what seems at first an elementary observation: The laws may have a symmetry that is not respected by all features of the world they apply to.

Let me illustrate this first with our social laws. Our laws apply equally to all people. We may regard this as a symmetry of the laws. Substitute any person for any other and you do not change the laws they must obey. All must pay taxes, all must not exceed the speed limit. But this equality or symmetry before the law need not and does not require that our circumstances be the same. Some of us are wealthier than others. Not all of us have cars, and in those that do, the tendencies to exceed the speed limit can differ quite a bit.

Moreover, in an ideal society we all start out with equal opportunities. This is unfortunately not actually the case, but if it were, we could speak of a symmetry in our initial opportunities. As life goes on, this initial symmetry goes away. By the time we turn twenty, we have very different opportunities. A few of us have the opportunity to be concert pianists, a few to be Olympic athletes.

We can describe this differentiation by saying that the initial equality is broken as time goes on. Physicists who speak of equality as a symmetry would say that the symmetry between us at birth is broken by the situations we encounter and the choices we make. In some cases, it would be hard to predict the way the symmetry will be broken. We know that it must break, but looking at a nursery full of infants we are hard-pressed to predict how. In cases like this, physicists say that the symmetry is *spontaneously broken*. By this we mean that it is necessary that the symmetry break, but exactly how it breaks is highly contingent. This spontaneous symmetry breaking is the second great principle that underlies the standard model of particle physics.

Here is another example from human life. As a faculty member, I've sometimes had occasion to go to receptions for new undergraduates. Watching them meet one another, it has occurred to me that over the next year some will become friends, others lovers, a few will even marry. At this first moment, when they encounter one another as strangers, there is a lot of symmetry in the room; many possible couples and bonds of friendship could be forged in this group. But of necessity the symmetry must be broken as the actual human relationships develop out of a much larger space of possible relationships. This, too, is an example of spontaneous symmetry breaking.

Much of the structure of the world, both social and physical, is a consequence of the requirement that the world, in its actuality, break symmetries present in the space of possibilities. An important feature of this requirement is the trade-off between symmetry and stability. The symmetric situation, in which we are all potentially friends and romantic partners, is unstable. In reality, we must make choices, and this leads to more stability. We trade the unstable freedom of potentiality for the stable experience of actuality.

The same is true in physics. A common example from physics is of a pencil balanced on its point. It is symmetric, in that while it is balanced on its point, one direction is as good as another. But it is unstable. When the pencil falls, as it inevitably must, it will fall randomly, in one direction or another, breaking the symmetry. Once it has fallen, it is stable, but it no longer manifests the symmetry — although the symmetry is still there in the underlying laws. The laws describe only the space of what possibly may happen; the actual world governed by those laws involves a choice of one realization from many possibilities.

This mechanism of spontaneous symmetry breaking can happen to the symmetries between the particles in nature. When it occurs for the symmetries that, by the gauge principle, give rise to the forces of nature, it leads to the differences in their properties. The forces become distinguished; they can have different ranges and different strengths. Before the symmetry breaks, all four fundamental forces have an infinite range, like electromagnetism, but afterward some will have a finite range, like the two nuclear forces. As noted, this is one of the most important discoveries of twentieth-century physics, because together with the gauge principle it allows us to unify fundamental forces that appear disparate.

The idea of combining spontaneous symmetry breaking with the gauge theories was invented by François Englert and Robert Brout in Brussels, in 1962, and independently a few months later by Peter Higgs of Edinburgh University. It should be called the EBH phenomenon, but it is unfortunately usually just called the Higgs phenomenon. (This is one of many examples in which something in science is named after the last person who discovered it, rather than the first.) The three of them also showed that there is a particle whose existence is a consequence of spontaneous symmetry breaking. This is called the *Higgs boson.*

A few years later, in 1967, Steven Weinberg and the Pakistani physicist Abdus Salam independently discovered that the combination of the gauge principle and spontaneous symmetry breaking could be used to construct a concrete theory that unified the electromagnetic and weak nuclear forces. The theory bears their name: the Weinberg-Salam model of the *electroweak force.* This was certainly a unification with consequences to be celebrated; it quickly led to predictions of novel phenomena that were successfully verified. It implies, for example, that there must be particles — analogous to the photon, which carries the electromagnetic force — to carry the weak nuclear force. There are three of them, called the W^+, W^-, and Z. All three have been found and exhibit exactly the properties predicted.

The use of spontaneous symmetry breaking in a fundamental theory was to have profound consequences, not just for the laws of nature but for the larger question of what a law of nature is. Before this, it was thought that the properties of the elementary particles are determined directly by eternally given laws of nature. But in a theory with spontaneous symmetry breaking, a new element enters, which is that the properties of the elementary particles depend in part on history and environment. The symmetry may break in different ways, depending on conditions like density and temperature. More generally, the properties of the elementary particles depend not just on the equations of the theory but on which solution to those equations applies to our universe.

This signals a departure from the usual reductionism, according to which the properties of the elementary particles are eternal and set by absolute law. It opens up the possibility that many — or even all — properties of the elementary particles are contingent and depend on which solution of the laws is chosen in our region of the universe or in our particular era. They could be different in different regions.[4] They could even change in time.

In spontaneous symmetry breaking, there is a physical quantity whose value signals that the symmetry is broken and how. This quantity is usually a field, called the Higgs field. The Weinberg-Salam model requires that the Higgs field exist and that it manifest itself as the new elementary particle called the Higgs boson, which carries the force associated with the Higgs field. Of all the predictions required by the unification of the electromagnetic and weak forces, only this one has not yet been verified. One difficulty is that the theory does not allow us to precisely predict the mass of the Higgs boson; it is one of the free constants that the theory asks us to set. There have been many experiments designed to find the Higgs boson, but all we know is that if it exists, its mass must be greater than about 120 times the mass of the proton. One of the main goals of future accelerator experiments is to find it.

In the early 1970s, the gauge principle was applied to the strong nuclear force, the force that binds the quarks, and it was found that a gauge field is responsible for that force, too. The resulting theory is called *quantum chromodynamics,* or QCD for short. (The word *chromo,* from the Greek for "color," refers to a fanciful designation used to refer to the fact that quarks come in three versions, which, for fun, are called colors.) QCD, too, has survived rigorous experimental test. Together with the Weinberg-Salam model, it is the basis of the standard model of elementary-particle physics.

The discovery that all three forces are expressions of a single unifying principle — the gauge principle — is the deepest accomplishment of theoretical particle physics to date. The people who did this are true heroes of science. The standard model is the result of decades of hard, often frustrating experimental and theoretical work by hundreds of people. It was completed in 1973, and it has held up for thirty years against a wide array of experiments. We physicists are justly proud of it.

But consider what happened next. All three forces were now understood to be expressions of the same principle, and it was obvious that they should be unified. To unify all the particles, however, you need a big symmetry that includes them all. You then apply the gauge principle, giving rise to the three forces. To distinguish all the particles and forces, you set it up so that any configuration of the system in which the symmetry is realized is unstable, while the stable configurations are asymmetrical. This is not hard to do because, as I discussed, symmetrical situations are often unstable in nature. Thus, the symmetry including all the particles together will be spontaneously broken. This can be done so that the three forces end up with the very properties they are observed to have.

The idea of grand unification was not only to bring the forces together but to invent a symmetry that turned quarks (the particles ruled by the strong force) into leptons (the particles ruled by the electroweak force), hence unifying the two basic kinds of particles, leaving just one kind of particle and one gauge field. The simplest candidate for this grand unification was known as the symmetry SU(5). The name is code for the five kinds of particles rearranged by the symmetry: the three colored quarks of each kind and two leptons (the electron and its neutrino). SU(5) not only unified quarks and leptons, it did so with unparalleled elegance, explaining concisely all that went into the standard model and making a necessity out of much that was previously arbitrary. SU(5) explained all the predictions of the standard model and, even better, made some new predictions.

One of these new predictions was that there had to be processes by which quarks can change into electrons and neutrinos, because in SU(5), quarks, electrons, and neutrinos are just different manifestations of the same underlying kind of particle. As we have seen, when two things are unified, there have to be new physical processes by which they can turn into one another. SU(5) indeed predicts such a process, which is similar to radioactive decay. This is a wonderful prediction, characteristic of grand unification. It is required by the theory and is unique to it.

The decay of a quark into electrons and neutrinos would have a visible consequence. A proton containing that quark is no longer a proton; it falls apart into simpler things. Thus, protons are no longer stable particles — they undergo a kind of radioactive decay. Of course, if this happened very often, our world would fall apart, as everything stable in it is made of protons. So if protons do decay, the rate must be very small. And that is exactly what the theory predicted: a rate of less than one such decay every 10^{33} years.

But even though this effect is extremely rare, it is within the range of a doable experiment, because there are an enormous number of protons in the world. So in SU(5) we had the best kind of unified theory, one in which there was a surprising consequence that didn't contradict what we knew and could be confirmed immediately. We could compensate for the extreme rareness of proton decay by building a huge tank and filling it with ultrapure water, on the chance that somewhere in the tank a proton would decay as often as a few times a year. You would have to shield the tank from cosmic rays, because these rays, which are constantly bombarding the earth, can blast protons apart. After that, because the decay of a proton produces a lot of energy, all you had to do was surround the tank with detectors and wait. Funds were raised, and huge tanks were built in mines deep underground. The results were impatiently awaited.

After some twenty-five years, we are still waiting. No protons have decayed. We have been waiting long enough to know that SU(5) grand unification is wrong. It's a beautiful idea, but one that nature seems not to have adopted.

Recently, I ran into a friend from graduate school — Edward Farhi, who has since become the director of the Center for Theoretical Physics at MIT. We hadn't had a serious conversation in perhaps twenty years, but we found we had a lot to talk about. We had both been reflecting on what had happened and not happened in particle physics during the twenty-five years since we got our PhDs. Eddie made important contributions to particle theory but now works mostly in the rapidly moving field of quantum computers. I asked him why, and he said that in quantum computing, unlike particle physics, we know what the principles are, we can work out the implications, and we can do experiments to test the predictions we make. He and I found ourselves trying to pinpoint when particle physics had ceased to be the fast-moving field that had excited us in graduate school. We both concluded that the turning point was the discovery that protons don't decay within the time predicted by the

SU(5) grand unified theory. "I would have bet my life — well, maybe not my life, but you know what I mean — that protons would decay," was how he put it. "SU(5) was such a beautiful theory, everything fit into it perfectly — then it turned out not to be true."

Indeed, it would be hard to underestimate the implications of this negative result. SU(5) is the most elegant way imaginable of unifying quarks with leptons, and it leads to a codification of the properties of the standard model in simple terms. Even after twenty-five years, I still find it stunning that SU(5) doesn't work.

Not that it's hard for us theorists to get around the current failure. You can just add a few more symmetries and particles to the theory, so that there are more constants to adjust. With more constants to adjust, you can then arrange for the decay of the proton to be as rare as you like. So you can easily make the theory safe from experimental failure.

That said, the damage is done. We lost the chance to observe a striking and unique prediction of a deep new idea. In its simplest version, grand unification made a prediction for the rate of proton decay. If grand unification is right but complicated, so that the proton-decay rate can be adjusted to anything we like, it has ceased to be explanatory. The hope was that unification would account for the values of the constants in the standard model. Instead, grand unification, if valid, introduces new constants that must be tuned by hand to hide effects that would disagree with experiment.

We see here an illustration of the general lesson described earlier. When you unify different particles and forces, you risk introducing instabilities into the world. This is because there are new interactions by which the unified particles can morph into each other. There is no way to avoid these instabilities; indeed, such processes are the very proof of unification. The only question is whether we are dealing with a good case — like the standard model, which made unambiguous predictions that were quickly confirmed — or with an unkind case, in which we have to fiddle with the theory to hide the consequences. This is the dilemma of modern theories of unification.

5

From Unification
to Superunification

THE FAILURE OF the first grand unified theories gave rise to a crisis in science that continues to this day. Before the 1970s, theory and experiment had developed hand in hand. New ideas were tested within a few years, ten at most. Each decade from the 1780s to the 1970s saw a major advance in our knowledge of the foundations of physics, and in each advance, theory complemented experiment, but since the end of the 1970s there has not been a single genuine breakthrough in our understanding of elementary-particle physics.

When a big idea fails, there are two ways to respond. You can lower the bar and retreat to incremental science, slowly probing the borders of knowledge with new theoretical and experimental techniques. Many particle physicists did this. The result is that the standard model has been very well tested experimentally. The most consequential finding of the last quarter century is that neutrinos have mass, but this revelation can be accommodated by a minor adjustment of the standard model. Apart from that, no modifications have been made.

The other way to respond to the failure of a big idea is to try for an even bigger one. At first a few theorists, then a growing number of them, took this road. It is a road we have had to take alone; so far, none of the new ideas have any support from experiment.

Of the big ideas that have been invented and studied during these years, the one that has gotten the most attention is called *supersymmetry*. If true, it will become as fundamental a part of our understanding of nature as relativity theory and the gauge principle.

We have seen that the big unifications find hidden connections between aspects of nature that were previously thought distinct. Space and time were originally two very different concepts; the special theory of relativity unified them. Geometry and gravity were once quite unrelated; general relativity unified them. But there are still two big classes of things that make up the world we inhabit: the particles (quarks, electrons, etc.) that comprise matter and the forces (or fields) by which they interact.

The gauge principle unifies three of the forces. But we're still left with those two distinct entities: particles and forces. Their unification was the goal of two previous attempts, the aether theory and the unified-field theory, and each failed. Supersymmetry is the third attempt.

Quantum theory says that particles are waves and waves are particles, but this does not really unify the particles with the forces. The reason is that in quantum theory there remain two broad classes of elementary objects. These are called *fermions* and *bosons.*

All the particles that make up matter, such as electrons, protons, and neutrinos, are fermions. All the forces consist of bosons. The photon is a boson, and so are the particles, like the W and Z particles, associated with the other gauge fields. The Higgs particle is also a boson. Supersymmetry offers a way to unify these two big classes of particles, the bosons and the fermions. And it does so in a very creative way, by proposing that every known particle has a heretofore unseen *superpartner.*

Roughly speaking, a supersymmetry is a process in which you can replace one of the fermions by a boson in some experiment without changing the probabilities of the various possible outcomes. This is tricky to do, because fermions and bosons have very different properties. Fermions must obey the *exclusion principle*, invented by Wolfgang Pauli in 1925, which says that two fermions cannot occupy the same quantum state. This is why all the electrons in an atom do not sit in the lowest orbital; once an electron is in a particular orbit, or quantum state, you cannot put another electron in the same state. The Pauli exclusion principle explains many properties of atoms and materials. Bosons, however, behave in the opposite way: They like to share states. When you put a photon into a certain quantum state, you make it more likely that another photon will find its way to that same state. This affinity explains many properties of fields, like the electromagnetic field.

So it seemed at first crazy that you could invent a theory in which you could replace a boson with a fermion and still get a stable world. But plausible or not, four Russians found that they could write down a consistent theory with just such a symmetry, which we now call supersymmetry. They were Evgeny Likhtman and Yuri Golfand in 1971 and Vladimir Akulov and Dmitri Volkov in 1972.

In those days, scientists in the West were fairly out of touch with scientists in the Soviet Union. Soviet scientists were only rarely allowed to travel, and they were discouraged from publishing in non-Soviet journals. Most Western physicists did not read the translations of Soviet journals, so there were several discoveries made in the U.S.S.R. that went unappreciated in the West. The discovery of supersymmetry was one of them.

So supersymmetric theories were invented twice more. In 1973, several kinds were discovered by two European physicists, Julius Wess and Bruno Zumino. Unlike that of the Russians, their work was noticed, and the ideas were quickly developed. One of their new theories was an extension of electromagnetism in which the photon was unified with a particle much like a neutrino. The other discovery of supersymmetry is connected with string theory, and we'll explore it in more detail later.

Could supersymmetry be true? Not in its initial form, which posited that for each fermion there is a boson with the same mass and charge. This means there must be a boson with the same mass and charge as the electron. This particle, were it to exist, would be called a *selectron*, for superelectron. But if it existed, we would have already seen it in an accelerator.

This problem can be fixed, however, by applying the idea of spontaneous symmetry breaking to supersymmetry. The result is straightforward. The selectron gets a large mass, so it becomes much heavier than the electron. By adjusting free constants of the theory — of which it turns out to have many — you can make the selectron as heavy as you like. But there is an upper limit to how massive a particle any given particle accelerator can produce. Thus you can explain why it has not yet been seen in any existing particle accelerator. This is precisely what was done.

Notice that this story has a similar arc to others we have described. Someone posits a new unification. There are big consequences for experiment. Unfortunately, experiment disagrees. Scientists then complicate the theory, in a way that incorporates several adjustable constants. Finally, they adjust those constants to hide the missing predicted phenomena, thus explaining why the unification, if true, has not resulted in any observations. But such maneuvering makes the theory hard to falsify, because you can always explain away any negative result by adjusting the constants.

The story of supersymmetry is one in which, from the beginning, the game has been to hide the consequences of the unification. This does not mean that supersymmetry isn't valid, but it does explain why, even after more than three decades of intensive development, there are still no unambiguous testable predictions.

I can only imagine how Wess, Zumino, and Akulov (the only one of their Russian colleagues still alive) must feel. They may have made the most important discovery of their generation. Or they may simply have invented a theoretical toy that has nothing to do with nature. So far, there is no evidence either way. In the last thirty years, the first thing done with every new elementary-particle accelerator that has come on line has been to look for the particles that supersymmetry predicts. None has been found. The constants are just adjusted upward, and we wait again for the next experiment.

Today that means keeping an eye on the Large Hadron Collider (or LHC), presently under construction at CERN. If all goes according to plan, it should turn on in 2007. There is great hope among particle physicists that this machine will rescue us from the crisis. First of all, we want the LHC to see the Higgs particle, the massive boson responsible for carrying the Higgs field. If it doesn't, we will be in big trouble.

But the idea with the most at stake is supersymmetry. If the LHC sees supersymmetry, there will certainly be Nobel Prizes for its inventors. If not, there will be dunce caps — not for them, for there is no shame in inventing a new kind of theory, but for all those of my generation who have spent their careers extending that theory.

So many hopes are riding on the LHC because what it finds will also tell us a lot about one of the five key problems mentioned in chapter 1: how to explain the values of the free constants of the standard model. To see why, we need to understand one very striking feature of these values, which is that they are either very large or very small. One example is the wildly differing strengths of the forces. The electrical repulsion between two protons is stronger than their gravitational attraction by a huge factor, around 10^{38}. There are also huge differences in the masses of the particles. For example, the electron has $1/1{,}800$ the mass of a proton. And the Higgs boson, if it exists, has a mass of at least 120 times that of the proton.

A way to summarize the data is to say that particle physics seems hierarchical rather than democratic. The four forces span a large range of strengths, forming a hierarchy from strong to weak, which is to say from nuclear physics to gravity. The various masses in physics also form a hierarchy. At the top is the Planck mass, which is the energy (recall that mass and energy are really the same thing) at which quantum gravity effects will become important. Perhaps ten thousand times lighter than the Planck mass is the scale at which the difference between electromagnetism and the nuclear forces should be seen. Experiments conducted at that energy, which is called *the unification scale,* will see not three forces but one single force. Moving down the hierarchy, 10^{-16} times the Planck scale is a TeV (for tera-electron volt, or 10^{12} electron volts), the energy at which the unification of the weak and electromagnetic forces takes place. This is called *the weak interaction scale.* This is the region in which we should see the Higgs boson, and it is also where many theorists expect to see supersymmetry. The LHC is being built to probe the physics at this scale. A proton mass is $1/1{,}000$ of that, another $1/1{,}000$ brings us down to the electron, and perhaps $1/1{,}000{,}000$ of that is the neutrino. Then, way down at the bottom, is the vacuum energy, which exists throughout space even in the absence of matter.

This makes a beautiful but puzzling picture. Why is nature so hierarchical? Why is the difference between the strength of the strongest and weakest force so huge? Why are the masses of protons and electrons so tiny compared with the Planck mass or the unification scale? This problem is generally referred to as *the hierarchy problem,* and we hope the LHC will shed light on it.

So what exactly should we see at the LHC? This has been the central question of particle physics since the triumph of the standard model in the early 1970s. Theorists have had three decades to prepare for the day the LHC goes on line. Are we ready? Embarrassingly, the answer is no.

Were we ready, we would have a compelling theoretical prediction for what the LHC will see, and we'd simply be awaiting confirmation. Given everything we do know about particle physics, it is surprising that thousands of the smartest people on the planet have been unable to come up with a compelling guess as to what the next great experimental leap will reveal. But apart from the hope that the Higgs boson will be seen, we have no clear, unambiguous prediction.

You might think that in the absence of consensus, there would be at least a few rival theories making such a prediction. But the reality is far messier. We have several different unification proposals on the table. All potentially work, to some extent, but none has emerged as uniquely simpler or more explanatory than the others. None yet has the ring of truth. To explain why thirty years have not sufficed to put our theoretical house in order, we need to look more closely at the hierarchy problem. Why is there such an enormous range of masses and other constants?

The hierarchy problem contains two challenges. The first is to determine what sets the constants, what makes ratios large. The second is how they stay there. This stability is puzzling, because quantum mechanics has a strange tendency to pull all the masses together toward the value of the Planck mass. We don't need to explore why here, but the result is as if some of the dials we use to tune the constants were connected by rubber bands that are steadily tightening.

As a consequence, we can keep the large ratios in the standard model, but to do so we have to pick the constants precisely. The larger the ratios we want the actual masses to have, the more finely we theorists have to tune the intrinsic masses (the masses absent quantum effects) to keep them apart. Just how finely depends on the kinds of particles involved.

The masses of the gauge bosons are not much of a problem; the symmetry basically prevents the rubber bands from pulling on their masses. And both before and after quantum effects are taken into account, a photon, which is the boson that carries the electromagnetic field, has no mass at all, so it is not a problem either. Nor are the constituents of matter, the quarks and the leptons; the parts of their masses that come from quantum effects are proportional to their intrinsic masses. If the intrinsic masses are small, so are the total masses. We say, therefore, that the masses of the gauge bosons and the fermions are *protected.*

The problem lies with the unprotected particles, which in the standard model of particle physics means the Higgs and the Higgs alone. It turns out that to protect the mass of the Higgs from being pulled up to the Planck mass, we have to tune the constants of the standard model to the amazing precision of thirty-two decimal places. Any inaccuracy in any one of those thirty-two decimal places and the Higgs boson ends up much heavier than it is predicted to be.

The challenge is then to tame the Higgs — to bring it down to size, so to speak. Many of the big ideas that particle physicists have explored since 1975 aim to do just that.

One way to tame the Higgs is to propose that it is not an elementary particle at all. If it were made of particles that behaved less wildly, the problem could be eliminated There are several proposals for what the Higgs boson might be made of. The most elegant and sparse theory hypothesizes that Higgs bosons are bound states of very heavy quarks or leptons. Nothing new is added at all — no new particles and no parameters to tune. The theory just posits that heavy particles stick together in new ways. The only problem with this kind of theory is that it is hard to do the calculations required to check it and work out the consequences. It was beyond our technology to do so when first proposed in the 1960s, and it still is.

The next-most-elegant hypothesis is that the Higgs boson is made up of a new kind of quark, different from those that make up

protons and neutrons. Because this seemed at first a "technical" solution to the problem, these were called *techniquarks*. They are
bound together by a new kind of force, similar to the strong nuclear
force that binds quarks into protons and neutrons. Since the force
in quantum chromodynamics is sometimes called "color," the new
force is called, of course, *Technicolor.*

This idea is more amenable to calculation. The problem is that it
is hard to get this theory to agree with all aspects of the observations. But it's not impossible, since there are many variants. Most
have been ruled out; a few remain viable.

A third option is to make *all* the elementary particles into composite particles. This idea was pursued by a few people in the late
1970s. It was a natural thing to try: If protons and neutrons are made
of quarks, why stop there? Perhaps there is a further level of structure, where quarks, electrons, neutrinos, and perhaps even the Higgs
and the gauge bosons are seen to be made of particles that are even
more fundamental and that we might call *preons.* Such theories
worked very elegantly. Experiment had by then given us evidence
for the forty-five fundamental fermions, and all could be put together from combinations of just two kinds of preons.

Moreover, these preon models explained some features observed
in nature but unexplained in the standard model. For example, the
quarks have two properties — color and charge — that seem unrelated. Each kind of quark comes in three versions called colors. This
triplication provides the symmetry required for the gauge theory.
But why three colors? Why not two, or four? Each quark also has an
electric charge, and these come in units that are $1/3$ and $2/3$ of the
electron's charge. The number 3 occurs in each case, which suggests
that these two properties, color and charge, could have a common
origin. Neither the standard model nor, to my knowledge, string
theory addresses this coincidence, but it is explained very simply by
the preon model.

Unfortunately, there were major questions that the preon theories were not able to answer. These have to do with the unknown
force that must bind the preons together into the particles we observe. The challenge was to keep the observed particles as small as
they are while keeping them very light. Because preon theorists

couldn't solve this problem, preon models were dead by 1980. I've talked recently with well-known physicists who got their PhDs after this and have never even heard of them.

So, altogether, the attempts to make the Higgs boson a composite were not convincing. It seemed for a time that we theorists were running out of options. If the Higgs boson is elementary, then how can its properties be tamed?

One way to limit the freedom of a particle is to tie its behavior to another particle whose behavior is constrained. We know that the gauge bosons and the fermions are protected; their masses do not behave wildly. Could there be a symmetry that ties the Higgs to a particle whose mass is protected? If we could do that, perhaps the Higgs would be tamed at last. The only symmetry known to do this is supersymmetry, because supersymmetry relates fermions to bosons; hence, in a supersymmetric theory there will be a fermion that partners with the Higgs, called the Higgsino. (In supersymmetry-theory convention, the superpartners of fermions begin with an "s," like the selectron, while the superpartners of bosons end in "ino.") Because it is a fermion, the mass of the Higgsino will be protected from quantum weight gain. Well, supersymmetry tells us that two partners have the same mass. So the mass of the Higgs must be protected too.

This idea might well explain why the Higgs mass is low compared to the Planck mass. As stated, this idea is certainly elegant — but in practice it is complicated.

First of all, a theory cannot be partly supersymmetric. If one particle has a superpartner, they all must. Thus, each quark comes with a bosonic partner, a *squark*. The photon is partnered with a new fermion, the *photino*. The interactions are then tuned so that if all quarks are replaced by squarks at the same time as we replace all photons by photinos, the probabilities of the various possible outcomes are unchanged.

Of course, there is a simpler possibility. Couldn't two particles that we have already observed be partnered? Perhaps the photon and the neutrino go together? Or the Higgs and the electron? The discovery of a new relationship among known particles would certainly be elegant — and convincing.

Unfortunately, no one has ever successfully postulated a supersymmetry holding between two known particles. Instead, in all the supersymmetric theories the numbers of particles are at least doubled. A new superpartner is simply postulated to go along with each known particle. Not only are there squarks and sleptons and photinos, there are also sneutrinos to partner the neutrinos, Higgsinos with the Higgs, and gravitinos to go with the gravitons. Two by two, a regular Noah's ark of particles. Sooner or later, tangled in the web of new snames and naminos, you begin to feel like Sbozo the clown. Or Bozo the clownino. Or swhatever.

For better or worse, nature is not like this. As noted, no experiment has ever produced evidence for a selectron. There appear to be, so far, no squarks, no sleptons, and no sneutrinos. The world contains huge numbers of photons (more than a billion for every proton), but no one has ever seen even a single photino.

The solution to this is to posit that supersymmetry is spontaneously broken. We discussed in chapter 4 how a symmetry is spontaneously broken. This spontaneous breaking can be extended to supersymmetry. Theories can be constructed that describe worlds in which the forces are supersymmetric but where those laws are cleverly adjusted so that the lowest energy state — that is, the state at which symmetry disappears — is not supersymmetric. As a result, the supersymmetric partner of a particle need not have the same mass that the particle has.

This makes for an ugly theory. To break the symmetry, we have to add still more particles, analogous to the Higgs. They also need superpartners. There are still more free constants, which can be adjusted to describe their properties. All the constants of the theory then have to be adjusted so that all of these new particles are too heavy to be observed.

Doing this to the standard model of elementary-particle physics, with no additional assumptions, results in a contraption called the *minimally supersymmetric standard model,* or MSSM. As noted in chapter 1, the original standard model has about 20 free constants we have to adjust by hand to get predictions that agree with experiment. The MSSM adds 105 more free constants. The theorist is at liberty to adjust them all to ensure that the theory agrees with experiments. If this theory is right, then God is a techno-geek. He is the kind of guy or gal who likes a music system with as many dials as possible or a sailboat with 16 different lines to adjust the shape of each sail.

Of course, nature may be like this. The theory does have the potential to solve the fine-tuning problem. So what you get for increasing the number of dials from 20 to 125 is that none of the new dials have to be as finely tuned as the old dials. Still, with so many dials to adjust, the theory is difficult for experimentalists to prove or disprove.

There are many settings of the dials for which the supersymmetry is broken and each particle has a mass different from that of its superpartner. To hide all the missing better halves, we have to tune the dials in such a way that the missing particles all end up with a lot more mass than the ones we see. You have to get this right, for if the theory predicted that the squarks were lighter than the quarks, we would be in trouble. Not to worry. There turn out to be many different ways to tune the dials to ensure that all the particles we don't see are so heavy that they're as yet unseeable.

If the fine-tuning is to be explained, then the theory has to give an explanation for why the Higgs boson has the large mass we think it has. As noted, there is not an exact prediction for the Higgs mass even in the standard model, but it has to be more than about 120 times the mass of a proton. To predict this, a supersymmetric theory must be tuned so that at this scale the supersymmetry is restored. This means that the missing superpartners should have masses at about this scale, and if so, the LHC should see them.

Many theorists expect that this is what the LHC will see — lots of new particles that can be interpreted as missing superpartners. If the LHC does so, it certainly will be a triumph for the last thirty years of theoretical physics. However, I remind you that there are no clear predictions. Even if the MSSM is true, there are many different ways to tune its 125 parameters to agree with what is known at present. These lead to at least a dozen very distinct scenarios, which make quite different predictions about exactly what the LHC will see.

There are further worries. Suppose the LHC produces a new particle. Given that the supersymmetric theory comes in many different scenarios, it is possible that even if supersymmetry is wrong, it could still be adjusted to agree with the first observations from the

LHC. To confirm supersymmetry, much more is needed. We'll have to discover many new particles and explain them. And they may not all be superpartners of particles we know about. A new particle could be a superpartner of yet another new particle, still unseen.

The only unimpeachable way to prove supersymmetry true will be to show that there really is a symmetry — which is to say that the probabilities for the various possible outcomes of experiments don't change (or change in certain very restricted ways) when we substitute one particle for its superpartner. But this is something that will not be easy for the LHC, at least initially. So even in the best of circumstances, it will be many more years before we know whether supersymmetry is the right explanation for the fine-tuning problem.

Meanwhile, a great many theorists appear to believe in supersymmetry. And there are a few good reasons for thinking it is an improvement on older ideas of unification. First, the Higgs boson, if not pointlike, does not appear to be very large. This favors supersymmetry while ruling out some (though not all) Technicolor theories. There is also an argument that comes from the idea of grand unification. As we discussed earlier, experiments done at the unification scale should not be able to distinguish between electromagnetism and the nuclear forces. The standard model predicts this kind of unified scale but requires slight adjustments. The supersymmetric version gives unification more directly.

Supersymmetry is certainly a very compelling theoretical idea. The idea of a unification of forces and matter offers a resolution of the deepest duality in fundamental physics. No wonder so many theorists cannot imagine that the world is not supersymmetric.

At the same time, some physicists do worry that supersymmetry, if real, should have already been seen in experiment. Here is a fairly typical quote, from an introduction to a recent paper: "Another problem comes from the fact that LEP II [the Large Electron-Positron accelerator, also at CERN] did not discover any superparticles or the Higgs boson."[1] Paul Frampton, a distinguished theorist at the University of North Carolina, recently wrote me that,

> One general observation I have made over the last decade or more is that the majority of researchers (there are a few exceptions) working on the phenomenology of TeV scale supersymmetry breaking think

that the probability that TeV scale supersymmetry will show up in experiment is much less than 50 percent, an estimate of 5 percent being quite typical.[2]

My own guess, for what it's worth, is that (at least in the form so far studied) supersymmetry will not explain the observations at the LHC. In any case, supersymmetry is decidable by experiment, and whatever our aesthetic preferences, we will all be thrilled to have an answer to the question of whether or not it is a true picture of nature.

But even if supersymmetry were detected, it would not in itself be a solution to any of the five big problems I listed in chapter 1. The constants of the standard model would not be explained, because the MSSM has many *more* free constants. The possible choices for a quantum theory of gravity would not be narrowed, because the leading theories are all consistent with the world's being supersymmetric. It may be that the dark matter is made up of superpartners, but we would need this to be confirmed directly.

The reason for this larger inadequacy is that while supersymmetric theories have much more symmetry, they are not simpler. They are in fact much more complicated than theories with less symmetry. They do not decrease the number of free constants — they increase them, drastically. And they fail to unify any two things we already know about. Supersymmetry would be absolutely compelling — as compelling as Maxwell's unification of electricity and magnetism — if it uncovered a deep commonality between two known things. If the photon and electron turned out to be superpartners, say, or even the neutrino and the Higgs, it would be fantastic.

But this is not what any of the supersymmetry theories have done. Instead, they posit a whole new set of particles and make each particle symmetric with either a known particle or another unknown particle. This kind of theoretical success is far too easy. To invent a whole new world of the unknown and then make a theory with many parameters — parameters that can be tuned to hide all the new stuff — is not very impressive, even if it's technically challenging to pull off. It is the kind of theorizing that can't fail, because any disagreement with present data can be eliminated by tweaking some constants. It can fail only when confronted with experiment.

Of course, none of this means that supersymmetry is not real. It may be, and if it is, there is a chance it will be discovered in the next few years, at the LHC. But the fact that supersymmetry does not do all that we hoped suggests that its proponents are sitting way out on a limb, far from the sturdy trunk of empirical science. Perhaps that is the cost of looking to drill, as Einstein said, where the wood is thin.

6

Quantum Gravity:
The Fork in the Road

WHILE MOST physicists were ignoring gravity, a few brave souls in the 1930s began to think about reconciling it with the rapidly developing quantum theory. For more than half a century, no more than a handful of pioneers would work on quantum gravity, and few would pay them any attention. But the problem of quantum gravity could not be ignored forever. Of the five questions I described in the first chapter, it is the one that cannot go unsolved. Unlike the others, it seeks nothing less than the language in which the laws of nature are written. To try to solve any of the other problems without solving this one would be like trying to negotiate a contract in a country without law.

The search for quantum gravity is a true quest. The pioneers were explorers in a new landscape of ideas and possible worlds. Now there are more of us, and some of the landscape has been well mapped. Some trails were explored only to lead to dead ends. And while some are still being blazed and a few are even becoming crowded, we cannot yet say that the problem is solved.

Much of this book was written in 2005, the centenary of Einstein's first great achievements. The year was full of conferences and events to mark the anniversary. It was as good an excuse as any to draw attention to physics, but it was not without irony. Some of Einstein's discoveries were so radical that even now they are insufficiently appreciated by many theoretical physicists; chief among these is the understanding he achieved of space and time in general relativity.

The main lesson of general relativity is that the geometry of space is not fixed. It evolves dynamically, changing in time as matter moves about. There are even waves — gravitational waves — that travel through the geometry of space. Until Einstein, the laws of Euclidean geometry we learned in school were seen as eternal laws: It always was and always would be true that the angles of a triangle add up to 180 degrees. But in general relativity the angles of a triangle can add up to anything, because the geometry of space can curve.

This doesn't mean that there is some other fixed geometry that characterizes space — that space is like a sphere, or a saddle, instead of a plane. The point is that the geometry can be anything at all, because it evolves in time, responding to matter and force. Rather than a law stating what the geometry is, there is a law that governs how the geometry changes — just as Newton's laws tell us not where objects are but how they move, by specifying what effects force has on their motion.

Before Einstein, geometry was thought to be part of the laws. Einstein revealed that the geometry of space is evolving in time, according to other, deeper laws.

It is important to absorb this point completely. *The geometry of space is not part of the laws of nature.* There is therefore nothing in those laws that specifies what the geometry of space is. Thus, before solving the equations of Einstein's general theory of relativity, you don't have any idea what the geometry of space is. You find out only after you solve the equations.

This means that the laws of nature have to be expressed in a form that does not assume that space has any fixed geometry. This is the core of Einstein's lesson. We encapsulate it in a principle we described earlier, which is *background independence*. The principle states that the laws of nature can be specified completely without making any prior assumption about the geometry of space. In the old picture, in which geometry was fixed, it could be thought of as part of the background, the unchanging stage on which the pageant of nature unfolds. To say that the laws of physics are background-independent means that the geometry of space is not fixed but evolves. Space and time emerge from the laws rather than providing an arena in which things happen.

Another aspect of background independence is that there is no preferred time. General relativity describes the history of the world most fundamentally in terms of events and relationships between them. The principal relationships have to do with causality; one event may be in the chain of causes leading to another event. From this point of view, space is a secondary concept. The concept of space is in fact entirely dependent on the notion of time. Given a clock, we can think of all the events that are simultaneous with the clock striking noon. These make up space.

An important aspect of the general theory of relativity is that there is no preferred way to keep time. Any sort of clock will do, as long as it shows causes preceding effects. But because the definition of space depends on time, there are as many different definitions of space as there are of time. Just above, I spoke about the geometry of space evolving in time. That holds not for a single universal notion of time but for every possible notion of time. How all this works is part of the intricate beauty of Einstein's general theory of relativity. For our purposes, it will be enough to remember that the equations of that theory tell us how the geometry of space evolves in time not just for one but for any possible definition of time.

Actually, background independence means even more than this. There are other aspects of nature that are fixed in the usual expressions of the laws of physics. But perhaps they shouldn't be. For example, the fact that there are only three dimensions of space is part of the background. Might there be a deeper theory in which we don't have to make any prior assumption about the number of spatial dimensions? In such a theory, the three dimensions might come out as the solution to some dynamical law. Perhaps, in such a theory, the number of spatial dimensions could even change in time. If we could invent such a theory, it might explain to us why our universe has three dimensions. This would constitute progress, for something that previously was simply assumed would finally be explained.

So the idea of background independence in its broadest terms is a piece of wisdom about how to do physics: Make better theories in which things that are now assumed are explained, by allowing such things to evolve subject to some new law. Einstein's general theory of relativity did precisely that for the geometry of space.

The key question for a quantum theory of gravity is then the following: Can we extend to quantum theory the principle that space has no fixed geometry? That is, can we make quantum theory background-independent, at least with regard to the geometry of space? If we can do this, we will automatically merge gravity and quantum theory, because gravity is already understood to be an aspect of dynamical spacetime geometry.

There are then two approaches to merging gravity and quantum theory: those that achieve background independence and those that do not. The field of quantum gravity split along these lines all the way back in the 1930s, although most approaches studied today are background-independent. The one exception is the approach that most of today's physicists study — string theory.

How it came to be that the highest achievement of the most famous scientist of the twentieth century has been virtually ignored by most of those clamoring to follow in his footsteps is one of the strangest stories in the history of science. But it is a story that must be told here, as it is central to the questions I raised in the Introduction. Indeed, you might wonder, given that Einstein's general theory of relativity is so well accepted, why anyone would try to develop a new theory that did not take on board its central tenet. The answer is a story, and like many stories told in this book, it began with Einstein.

Already in 1916, Einstein realized that there were gravitational waves and that they carried energy. He noticed right away that consistency with atomic physics would require that the energy carried by gravitational waves be described in terms of quantum theory. In the very first paper ever written on gravitational waves, Einstein said that "it appears that the quantum theory would have to modify not only Maxwell's theory of electrodynamics but also the new theory of gravitation."[1]

Nevertheless, whereas Einstein was the first to state the problem of quantum gravity, his deepest insight has been ignored by most of those who have since worked on it. How could this be?

There *is* a reason, and it is that no one knew at the time how to go about directly applying the then developing quantum theory to general relativity. Instead, progress turned out to be possible by an indirect route. Those who wanted to apply quantum mechanics to general relativity faced two challenges. Besides background independence, they had to grapple with the fact that general relativity is a field theory. There are an infinite number of possibilities for the geometry of space and hence an infinite number of variables.

As I described in chapter 4, as soon as quantum mechanics was completely formulated, physicists began to apply it to field theories, such as the electromagnetic field. These are formulated in a fixed-spacetime background, so the issue of background independence does not arise. But they gave physicists experience with handling the problem of an infinite number of variables.

The first big success of quantum field theory was QED, the unification of Maxwell's theory of electromagnetism with quantum theory. It is remarkable that in their first paper on QED, in 1929, Werner Heisenberg and Wolfgang Pauli, two of the founders of quantum mechanics, were already contemplating extending their work to quantum gravity. They apparently felt it would not be too hard, because they write that the "quantization of the gravitational field, which appears to be necessary for physical reasons, may be carried out without any new difficulties by means of a formalism fully analogous to that applied here."[2]

More than seventy-five years later, we can only marvel at the extent to which two such brilliant people underestimated the difficulty of the problem. What could they have been thinking? Well, I know, because many people have since had the same thought, and the dead end it leads to has been thoroughly explored.

What Heisenberg and Pauli were thinking was that when gravitational waves are very weak, they can be seen as tiny ripples disturbing a fixed geometry. If you drop a stone into a pond on a still morning, it causes tiny ripples that barely disturb the flat surface of the water, so it is easy to think that the ripples move on a fixed background given by that surface. But when water waves are strong and turbulent, as near a beach on a stormy day, it makes no sense to see them as disturbances of something fixed.

General relativity predicts that there are regions of the universe where the geometry of spacetime evolves turbulently, like waves crashing on a beach. But Heisenberg and Pauli thought it would be simpler to first study cases in which the gravitational waves are extremely weak and can be seen as tiny ripples on a fixed background. This allowed them to apply the same methods they had developed to study quantum electromagnetic fields moving on a fixed background of spacetime. And in fact it was not difficult to apply quantum mechanics to very weak gravitational waves moving freely. The result was that each gravitational wave could be seen quantum mechanically, as a particle called the graviton — analogous to the photon, which is the quantum of the electromagnetic field. But at the next step, they faced a big problem, because gravitational waves interact with each other. They interact with anything that has energy, and they themselves carry energy. This problem does not occur with electromagnetic waves, because though photons interact with electric and magnetic charges, they are not themselves charged, so they go right through one another. This important difference between the two kinds of waves is what Heisenberg and Pauli missed.

Describing the self-interaction of gravitons consistently turned out to be a tough nut to crack. We now understand that the failure to solve this problem is a consequence of not taking Einstein's principle of background independence seriously. Once the gravitational waves interact with one another, they can no longer be seen as moving on a fixed background. They *change* the background as they travel.

A few people already understood this in the 1930s. Probably the first PhD thesis ever written on the problem of quantum gravity was the 1935 dissertation of the Russian physicist Matvei Petrovich Bronstein. Those who recall him think of him as one of the two most brilliant Soviet physicists of his generation. He wrote in a 1936 paper that "the elimination of the logical inconsistencies [requires] rejection of our ordinary concepts of space and time, modifying them by some much deeper and nonevident concepts." Then he quoted a German proverb, "Let him who doubt it pay a *Thaler.*"[3] Bronstein's view was championed by a brilliant young French physicist, Jacques Solomon.

By now almost everyone who thinks seriously about quantum gravity agrees with Bronstein, but it has taken seventy years. One reason is that even such brilliant minds as Bronstein and Solomon could not escape the insanity of their time. A year after Bronstein wrote the paper I just quoted, he was arrested by the NKVD, and he was executed by a firing squad on February 18, 1938. Solomon became a member of the French Resistance and was killed by the Germans on May 23, 1942. Their ideas were lost to history. I have worked on the problem of quantum gravity all my life and I learned of them only while finishing this book.

The work of Bronstein was forgotten, and most physicists returned to the study of quantum field theory. As I described in chapter 4, it took until the late 1940s for QED to be developed. This success then inspired a few people to take up again the challenge of unifying gravity with quantum theory. Right away, two opposing camps sprang up. One of them followed Bronstein in taking the background independence of general relativity seriously. The other ignored background independence and followed Heisenberg and Pauli's route in their efforts to apply quantum theory to gravitational waves seen as moving on a fixed background.

Since background independence is one of the principles of general relativity, it would seem sensible to incorporate it into attempts to unify that theory with quantum theory. But as it turned out, things were not so simple. A few people — like the British physicist P.A.M. Dirac, and Peter Bergmann, a German who had begun his career as an assistant to Einstein in Princeton — did attempt to construct a background-independent theory of quantum gravity. They found it an arduous task. Such attempts did not bear fruit until the mid 1980s, but since then there has been a lot of progress in understanding quantum gravity from a background-independent point of view. Most quantum-gravity theorists now work on one of several background-independent approaches. We'll return to these later in the book, for they constitute the most important alternatives to string theory.

But none of these promising signs were apparent when people started along the quantum-gravity road in the 1950s. The limited progress made with background-independent methods looked puny, compared with the great strides that were being made in QED. So until the late 1980s, most people took the other route, which was to attempt to apply the methods of QED to general relativity. This was perhaps understandable. After formulating QED, people knew a lot about background-dependent quantum theories, but no one knew anything about what a background-independent quantum theory might look like, if it existed at all.

Since this was the route that led to string theory, it is worth retracing. Because the work from the 1930s had been forgotten, it had to be rediscovered. The theory of gravitons was worked out again in a PhD thesis by Bryce DeWitt, who was a student of Julian Schwinger's at Harvard in the late 1940s. For this and his many discoveries that followed, we regard DeWitt as one of the founders of the theory of quantum gravity.

But, as noted, a graviton theory was not enough. The graviton theory was fine as long as the gravitons just moved through space, but if that's all they did, there was no gravity, and certainly no dynamical or curved geometry. So this was not a unification of general relativity or gravity with quantum theory, it was just a unification of weak gravitational waves with quantum theory. The problems with the theory of gravitons reemerged in the early 1950s, as soon as people began again to study how they might interact with one another. From then until the early 1980s, a lot of work was expended on this self-interaction problem to keep it from contradicting the principles of quantum theory. None of this work succeeded.

It might be useful to stop and think about what this means in human terms. We are talking about thirty years of continual hard work, involving many complicated calculations. Imagine doing your income tax every day, all day, for a week, and still not getting the calculations to add up consistently. You have an error somewhere, but you can't find it. Now imagine a month spent like that. Can you stretch it to a year? Now imagine twenty years. Now imagine that there are a couple of dozen people around the world spending their time like this. Some are friends, some rivals. They all have their own schemes of how to make it work. Each scheme has so far failed, but if you were to try a slightly different approach, or combine two approaches, perhaps you might succeed. Once or twice a year, you go to an international conference, where you can present your new scheme to the other fanatics. This was the field of quantum gravity before 1984.

Richard Feynman was one of the first to attack this graviton problem. And why not? He had done such good work on QED, why shouldn't he apply the same methods to quantum gravity? So in the early 1960s he took a few months off from particle physics to see if he could quantize gravity. To give you a sense of what a backwater quantum gravity was back then, here is a letter Feynman wrote to his wife in 1962 about a meeting in Warsaw where he was presenting his work:

> I am not getting anything out of the meeting. I am learning nothing. Because there are no experiments, this field is not an active one, so few of the best men are doing work in it. The result is that there are hosts of dopes here . . . and it is not good for my blood pressure. Remind me not to come to any more gravity conferences![4]

Nevertheless, he made good progress and greatly clarified a technical issue having to do with probabilities, which are numbers between 0 and 1. Anything that is certain to happen is said to have probability 1, so the probability that *anything at all* happens is 1. Before Feynman did his work, no one could make the probabilities for various things to happen in quantum gravity add up to 1. Actually, Feynman made the probabilities add up only in the first level of approximation; a few years later, Bryce DeWitt figured out how to make it work at all levels. A year or so later, the same thing was figured out by two Russians, Ludwig Dmitrievich Faddeev and Victor Nicolaevich Popov. They couldn't have known of DeWitt's work, because the journal had sent his paper to an expert to review and the reviewer had taken more than a year to go over it. So, bit by bit, people solved some problems — but even if the probabilities could be made to add up to 1, the graviton theory as a whole never worked.

There were some side benefits of this work. The same method could be applied to the Yang-Mills theories that the standard model came to be based on. So by the time Steven Weinberg and Abdus Salam used those theories to unify the weak and electromagnetic interactions, the technology was in place to do real calculations. The results turned out better than in quantum gravity. As the Dutch theorist Gerard 't Hooft finally proved in 1971, the Yang-Mills theories were completely sensible as quantum theories. Indeed, like others before him, 't Hooft was studying Yang-Mills theory partly as a warm-up to an attack on the problem of quantum gravity. So the thirty years of work on quantum gravity was not a completely wasted effort; at least it enabled us to do particle physics sensibly.

But there was no saving quantum gravity. People tried all sorts of approximation methods. Since the standard model of particle physics made sense, many methods were developed to probe different features of it. One by one, each of these was tried on the problem of quantum gravity. Each failed. No matter how you organized the quantum theory of gravitational waves, as soon as you put in the fact that they interacted with one another, infinite quantities raised their head. No matter how you turned the problem around, the infinities could not be tamed. More years of work, more papers, more PhD theses, more presentations at conferences. Same situation. The bottom line is that by 1974 it was clear that a background-dependent approach to combining general relativity with quantum theory did not make sense.

There was, however, one thing that could be done with background-dependent methods. Rather than trying to quantize gravity and so understand the effect that quantum theory has on gravitational waves, we could turn the problem around and ask what effects gravity might have on quantum phenomena. To do that, we could study quantum particles moving in spacetimes where gravity is important, such as black holes or an expanding universe. Beginning in the 1960s, a lot of progress was made in this direction. It is an important direction, because some of the discoveries led to puzzles that later approaches, such as string theory, aimed to solve.

The first success was a prediction that when the gravitational field changed rapidly in time, elementary particles would be created. This idea could be applied to the early universe when it was rapidly expanding, and led to predictions that are used to this day in the study of the early universe.

The success of these calculations inspired a few physicists to try something harder, which was to study the effect that a black hole can have on quantum particles and fields. The challenge here is that whereas black holes have a region where the geometry evolves very rapidly, this region is hidden behind a horizon. The horizon is a sheet of light that is standing still. It marks the boundary of a region within which all light is pulled inward, toward the center of the black hole. Thus no light can escape from behind the horizon. From the outside, a black hole seems static, but just inside its horizon is a region where everything is pulled toward stronger and stronger gravitational fields. These end in a singularity, where everything is infinite and time stops.

The first crucial result connecting quantum theory to black holes was made in 1973 by Jacob Bekenstein, a young Israeli graduate student of John Archibald Wheeler's at Princeton. He made the amazing discovery that black holes have entropy. Entropy is a measure of disorder, and there is a famous law, called the second law of thermodynamics, holding that the entropy of a closed system can never decrease. Bekenstein worried that if he took a box filled with a hot gas — which would have a lot of entropy, because the motion of the gas molecules was random and disordered — and threw it into a black hole, the entropy of the universe would seem to decrease, because the gas could never be recovered. To save the second law, Bekenstein proposed that the black hole must itself have an entropy, which would increase when the box of gas fell in, so that the total entropy of the universe would never decrease. Working out some simple examples, he was able to show that the entropy of a back hole must be proportional to the area of the horizon that surrounds it.

This introduced a puzzle. Entropy is a measure of randomness, and random motion is heat. So shouldn't a black hole also have a temperature? A year later, in 1974, Stephen Hawking was able to show that a black hole must indeed have a temperature. He was also able to fix the precise proportionality between the area of a black hole's horizon and its entropy.

There is another aspect of the temperature of black holes predicted by Hawking, which will be important to us later, and it's that the temperature of a black hole is inversely proportional to its mass. This means that black holes behave differently from familiar objects. To get most things to heat up, you have to put energy into them. We *fuel a fire*. Black holes behave in the opposite way. If you put energy, or mass, in, you make the black hole more massive — and it cools off.[5]

This mystery has since challenged every attempt to make a quantum theory of gravity: How can we explain the entropy and temperature of black holes from first principles? Bekenstein and Hawking treated the black hole as a classical fixed background within which quantum particles moved, and their arguments were based on consistency with known laws. They did not describe the black hole as a quantum-mechanical system, because that can be done only in a quantum theory of spacetime. So the challenge for any quantum theory of gravity is to give us a deeper understanding of Bekenstein's entropy and Hawking's temperature.

The following year, Hawking found still another puzzle lurking in these results. Because a black hole has a temperature, it will radiate, like any hot body. But the radiation carries energy away from the black hole. Given enough time, all the mass in the black hole will turn into radiation. As it loses energy, the black hole gets lighter. And because of the property I just described, when it loses mass, it heats up, so it radiates faster and faster. At the end of this process, the black hole will have shrunk down to a Planck mass, and one needs a quantum theory of gravity to predict the final fate of the black hole.

But whatever its final fate, there appears to be a puzzle concerning the fate of information. During the life of a black hole, it will pull in huge amounts of matter, carrying huge amounts of intrinsic information. At the end, all that's left is a lot of hot radiation — which, being random, carries no information at all — and a tiny black hole. Did the information just disappear?

This is a puzzle for quantum gravity, because there is a law in quantum mechanics that says that information can never be destroyed. The quantum description of the world is supposed to be exact, and there is a result implying that when all the details are taken into account, no information can be lost. Hawking made a strong argument that a black hole that evaporates away loses information. This appears to contradict quantum theory, so he called this argument *the black-hole information paradox.* Any putative quantum theory of gravity needs to resolve it.

These discoveries of the 1970s were milestones on the way to a quantum theory of gravity. Since then, we have measured the success of an approach to quantum gravity partly by how well it answers the challenges posed by the entropy, temperature, and information loss in black holes.

At about this time, an idea was finally proposed about quantum

gravity that seemed to work, at least for a while. It involved applying the idea of supersymmetry to gravity. The result was supergravity.

I was present at one of the first presentations ever given of this
new theory. It was a conference in 1975 in Cincinnati on developments in general relativity. I was still an undergraduate at Hampshire College, but I went anyway, hoping to learn what people were
thinking about. I remember some beautiful lectures by Robert Ge-
roch, of the University of Chicago, who was then a star of the field,
on the mathematics of infinite spaces. He got a standing ovation for
one particularly elegant demonstration. Then, stuck in at the end
of the conference was a talk by a young postdoc named Peter van
Nieuwenhuizen. I recall that he was quite nervous. He began by saying that he was there to introduce a brand-new theory of gravity. He
had my full attention.

Van Nieuwenhuizen said that his new theory was based on supersymmetry, then a new idea unifying bosons and fermions. The
particles you get from quantizing gravitational waves are called
gravitons, and they are a type of boson. But for a system to have supersymmetry, it must have both bosons and fermions. General relativity has no fermions, so new fermions must be hypothesized to be
the superpartners of the gravitons. "Sgraviton" is not an easy word
to say, so they were called *gravitinos.*

Since the gravitino had never been seen, he said, we were free to
invent the laws it satisfies. For the theory to be symmetric under supersymmetry, the forces could not change when gravitinos were
substituted for gravitons. This put a lot of constraints on the laws,
and searching for solutions to those constraints required weeks of
painstaking calculations. Two teams of researchers finished nearly
simultaneously. Van Nieuwenhuizen was part of one of those teams;
the other included my advisor-to-be at Harvard, Stanley Deser, who
was working with one of the inventors of supersymmetry, Bruno
Zumino.

Van Nieuwenhuizen also spoke of a deeper way to think about
the theory. You begin by thinking about the symmetries of space
and time. The properties of ordinary space remain unchanged if
we ourselves rotate, because there is no preferred direction. They

also remain unchanged if we move from place to place, because the geometry of space is uniform. Thus, translations and rotations are symmetries of space. Recall that in chapter 4 I explained the gauge principle, which states that in some circumstances a symmetry can dictate the laws that the forces satisfy. You can apply this principle to the symmetries of space and time. The result is precisely Einstein's general theory of relativity. This is not how Einstein found his theory, but had Einstein not lived, it is how general relativity might have been found.

Van Nieuwenhuizen explained that supersymmetry can be seen as a deepening of the symmetries of space. This is because of a profound and beautiful property: If you change all the fermions into bosons and then change them back again, you get the same world you had before but with everything moved a little bit in space. I can't here explain why this is true, but it tells us that supersymmetry is in some way fundamentally connected to the geometry of space. As a consequence, if you apply the gauge principle to supersymmetry, the result is a theory of gravity — supergravity. Seen in this way, supergravity is a profound deepening of general relativity.

I was a newcomer to the field, dropping in on a conference. I didn't know anyone there, so I don't know what van Nieuwenhuizen's listeners thought about what he had to say, but I was deeply impressed. I went home thinking it was a good thing that the guy was so nervous, for if what he said was right, it would be really important.

During my first year of graduate school, I took a course with Stanley Deser, who lectured about the new theory of supergravity. I got interested and started to think about it, but I was puzzled. What did it mean? What was it trying to tell us? I made a new friend there, a classmate named Martin Rocek, and he got excited as well. He quickly hooked up with Peter van Nieuwenhuizen, who was at Stony Brook, and began collaborating with him and his students. Stony Brook was not far away, and Martin brought me along on one of his visits there. Things were taking off in a big way, and he wanted to give me a chance to get in on the ground floor.

It was like being offered one of the first jobs at Microsoft or Google. Rocek, van Nieuwenhuizen, and many of those I met through them have made brilliant careers out of supersymmetry and supergravity.

I'm sure that from their point of view, I acted like a fool and blew a brilliant opportunity.

For me (and for others, I'm sure), the merging of supersymmetry with a theory of space and time raised profound questions. I had learned general relativity from reading Einstein, and if I understood anything, it was how that theory merged gravity with the geometry of space and time. That idea was in my bones. Now I was being told that another deep aspect of nature was also unified with space and time — the fact that there are fermions and bosons. My friends told me this, and the equations said the same thing. But neither friends nor equations told me what it meant. I was missing the idea, the conception of the thing. Something in my understanding of space and time, of gravity and of what it meant to be a fermion or boson, should deepen as a result of this unification. It should not just be math — my very conception of nature should change.

But it didn't. What I found when I hung out with van Nieuwenhuizen's students was a group of smart, technically minded kids frantically doing calculations, day and night. What they were doing was inventing versions of supergravity. Each version had a larger set of symmetries than the last, unifying a larger family of particles. They were working toward an ultimate theory that would unify all particles and forces with space and time. This theory had only a technical name, the $N = 8$ *theory*, N being the number of different ways to mix up fermions and bosons. The first theory — the one that van Nieuwenhuizen and Deser had introduced me to — was the simplest, $N = 1$. Some people in Europe had made $N = 2$. The week I was at Stony Brook, the people there were advancing toward $N = 4$, on their way to $N = 8$.

They worked day and night, ordered food in, and put up with the tedium of the work with the giddy certainty that they were on to something new and world-changing. One of them told me that he was working as fast as he could because he was sure that when the word got out about how easy it was to make new theories, the field would be overrun. Indeed, if I recall right, that group did get $N = 4$, but they were scooped on $N = 8$.

What they were doing didn't seem easy to me. The calculations were mind-numbingly lengthy and tedious. They required complete precision: If one factor of 2 went missing somewhere, weeks of work might have to be thrown out. Each line of the calculation had dozens of terms. To make a line of a calculation fit on a page, they resorted to larger and larger pads of paper. Soon they were carrying around huge art pads, the biggest they could find. They covered each page in tiny, precise handwriting. Each pad represented months of work. The word "monastic" comes to mind. I was terrified. I stayed a week and fled.

For decades afterward, I had rather uncomfortable relations with Peter, Martin, and the others. It may be that I was considered a loser for fleeing when they offered me the opportunity to join them in launching supergravity. Had I joined up, I might have been well placed to become one of the leaders of string theory. What I did instead was to go off in my own direction, eventually helping to found a different approach to the problem of quantum gravity. That made matters even worse: I was not only a loser who had abandoned the true faith, I was a loser in danger of becoming a rival.

As I reflect on the scientific careers of the people I have known these last thirty years, it seems to me more and more that these career decisions hinge on character. Some people will happily jump on the next big thing, give it all they've got, and in this way make important contributions to fast-moving fields. Others just don't have the temperament to do this. Some people need to think through everything very carefully, and this takes time, as they get easily confused. It's not hard to feel superior to such people, until you remember that Einstein was one of them. In my experience, the truly shocking new ideas and innovations tend to come from such people. Still others — and I belong to this third group — just have to go their own way, and will flee fields for no better reason than that it offends them that some people are joining in because it feels good to be on the winning side. So I no longer get bothered when I disagree with what other people are doing, because I see that temperament pretty much determines what kind of science they will do. Luckily for science, the contributions of the whole range of types are needed. Those who do good science, I've come to think, do so because they choose problems that are suited to them.

In any case, I fled the Stony Brook supergravity group but I didn't lose interest in supergravity. On the contrary, I was more interested than ever. I was sure they were on to something, but the road they were taking was not one I could follow. I understood Einstein's general theory of relativity, which meant that I knew how to demonstrate every essential property of it in a page or less of concise and transparent work. It seemed to me that if you understood a theory, it shouldn't take weeks of calculations on an art pad to check its basic properties.

I teamed up with another graduate student — a friend of mine from Hampshire College, John Dell, who was at the University of Maryland. We wanted to understand more deeply how it was that supersymmetry was part of the geometry of space and time. He found some papers by a mathematician named Bertram Kostant on a new kind of geometry that extended the math Einstein used by adding new properties that seemed to behave a bit like fermions. We wrote the equations of general relativity in this new context, and out popped some of the equations of supergravity. We had our first scientific paper.

At about the same time, others developed an alternative approach to a geometry for supergravity called supergeometry. I felt then (and feel now) that their setup is clumsier than ours. It is much more complicated, but for certain things it works much better. It helped to simplify the calculations somewhat, and that was certainly appreciated. So supergeometry caught on, and our work was forgotten. John and I didn't care, because neither approach gave us what we were looking for. Whereas the math worked, it didn't lead to any conceptual leaps. To this day, I don't think anyone really understands what supersymmetry means, what it says fundamentally about nature — if it's true.

Many years removed, I think I can finally fully articulate what drove me away from supergravity in those early days. Having learned physics by studying Einstein in the original, I had obtained a sense of the kind of thinking that went into a revolutionary new unification of physics. What I expected was that a new unification would start from a deep principle, like the principle of inertia or the equivalence principle. You would gain from this a deep and surprising insight that two things you had once seen as unrelated were actually at root the same thing. Energy is mass. Motion and rest are indistinguishable. Acceleration and gravity are the same.

Supergravity was not doing this. Although it was indeed a proposal for a new unification, it was one that could be expressed, and checked, only in the context of mind-crushingly boring calculations. I could do the math, but this was not the way I had been taught to do science by my readings of Einstein and the other masters.

Another friend I made at that time was Kellogg Stelle, who was a few years older than I was and, like me, was a student of Stanley Deser. Together they were exploring the question of whether supergravity behaved better than general relativity when combined with quantum theory. Since there had still been no progress on the background-independent methods, they, like everyone else, used the background-dependent method that had failed so miserably when applied to general relativity. They were quickly able to see that it worked better when applied to supergravity. They checked the first place where an infinity had occurred in quantum general relativity and found instead a finite number.

This was good news: Supersymmetry really did improve the situation! But the elation didn't last long. It took Deser and Stelle only a few more months to convince themselves that infinities would abound in supergravity farther along. The actual calculations were too hard to do, even after months of work with art pads, but they found a way to test whether the results would ultimately be finite or infinite, and it turned out that all the answers more precise than the one they could check — which had turned out finite — would be infinite.

They were not yet done, however, for there were all the other forms of supergravity to be tested. Perhaps one of those would finally yield a consistent quantum theory. One by one, each form was studied. Each one was a bit more finite, so that you had to go out farther in the sequence of approximations before the test failed. While the calculations were all too hard to do, there seemed no reason for any answer to be other than infinite past that point. There was a bit of hope that the final theory, the famous $N = 8$, would be different. It had finally been constructed by heroic work carried out in Paris. But it, too, failed the test — though there are some still holding out hope for it.

Supergravity was and is a wonderful theory. But by itself it was not enough to solve the problem of quantum gravity.

Thus, by the early 1980s there had been no progress on making a theory of quantum gravity. Everything that had been tried, up to and including supergravity, had failed. As the gauge theories triumphed, the field of quantum gravity stagnated. Those few of us who insisted on worrying about quantum gravity felt like the high school dropout invited to watch his sister graduate from Harvard with simultaneous degrees in medicine, neurobiology, and the history of dance in ancient India.

If the failure of supergravity to lead to a good theory of quantum gravity depressed us, though, it was also liberating. All the easy things had been tried. For decades, we had attempted to make a theory by extending the methods of Feynman and his friends. There were now only two things to try: Give up methods based on a fixed background geometry, or give up on the idea that the things moving through the background geometry were particles. Both approaches were about to be explored, and both would yield — for the first time — dramatic successes on the road to quantum gravity.

II

A BRIEF HISTORY
OF STRING THEORY

7

Preparing
for a Revolution

SOMETIMES SCIENTIFIC PROGRESS stalls when we encounter a problem that just cannot be solved in the way we understand it. There is a missing element, a different sort of trick involved. No matter how hard we work, we won't find the answer until someone somehow stumbles on this missing link.

Perhaps the first time this happened was with eclipses. Given the drama of a sudden darkening of the sky, the first order of business for early astronomers must have been to find a way to predict these scary events. Beginning several thousand years ago, people began to keep records of observations of eclipses, along with the motions of the sun, moon, and planets. It didn't take them long to understand that the motion of the sun and moon is periodic; we have evidence that people knew that back in our cave-dwelling days. But eclipses were harder.

A few things would have been clear to early astronomers. Eclipses happen when the sun and the moon, which take different paths across the sky, meet each other. Their paths intersect at two places. For an eclipse to happen, the sun and moon must meet at one of these two points. So in order to predict eclipses, you have to keep track of the annual path of the sun and the monthly path of the moon. Simply follow the paths and note when the two bodies meet.

The implication is that there must be a pattern that repeats in some multiple of the twenty-nine-and-a-half-day lunar period.

But this simple idea doesn't work: Eclipses do not fall into a pattern governed exactly by the lunar month. We can easily imagine the generations of theorists who tried and failed to reconcile the motions of these two great bodies. It would have been as great a puzzle to them as reconciling general relativity and quantum theory is to us.

We do not know who realized that there was an element missing, but we owe whoever it was a great debt. We can imagine an astronomer, perhaps in Babylon or ancient Egypt, suddenly realizing that there were not just two periodic motions to consider but three. Perhaps it was a sage, who after decades of study knew the data by heart. Perhaps it was some young rebel, not yet brainwashed into thinking that you had to explain what was seen only in terms of observable objects. Whatever the case, this innovator uncovered a mysterious third oscillation in the data, occurring not once a month or once a year but approximately every eighteen and two-thirds years. It turns out that the points where the two paths cross on the sky are not fixed: They rotate as well, taking those eighteen-plus years to make a complete cycle.

The discovery of this third motion — the missing element — must have been one of the earliest triumphs of abstract thinking. We see two objects, the sun and the moon. Each has a period, known from earliest times. It took an act of imagination to see that something else was moving as well: the paths themselves. This was a profound step, because it required realizing that behind the motion you observe there are other motions whose existence can only be deduced. Just a few times since has science progressed by the discovery of such a missing element.

The idea that elementary particles are not pointlike particles but vibrations of strings may be another of these rare insights. It provides a plausible answer to several big problems of physics. If right, it is as profound a realization as the ancient discovery that the circles the planets travel on are themselves moving.

The invention of string theory has been called a scientific revolution, but it was a long time in the making. As in some political revolutions — but unlike scientific revolutions of the past — the string theory revolution was anticipated by a small vanguard, who toiled for years in relative isolation. They began in the late 1960s to explore what happened when the strongly interacting particles — that is, particles made from quarks, such as protons and neutrons, and thus governed by the strong nuclear force — scatter off one another. This is not one of the five problems, because it is now understood, at least in principle, in terms of the standard model. But before the standard model was invented, it was a central problem for elementary-particle theorists.

Besides protons and neutrons, there are a great many other particles made from quarks. These others are unstable; they are produced in accelerators by smashing a beam of protons at high energy into other protons. From the 1930s to the 1960s, we accumulated a lot of data about the various kinds of strongly interacting particles and what happened when two of them collided.

In 1968, a young Italian physicist named Gabriele Veneziano saw an interesting pattern in the data. He described the pattern by writing down a formula that described the probabilities for two particles to scatter from each other at different angles. Veneziano's formula fit some of the data remarkably.[1]

It caught the interest of some of his colleagues, in Europe and in the United States, who puzzled over it. By 1970 a few were able to interpret it in terms of a physical picture. According to this picture, particles could not be seen as points, which is how they had always been seen before. Instead, they were "stringlike," existing only in a single dimension, and they could be stretched, like rubber bands. When they gained energy, they stretched; when they gave up energy, they contracted — also just like rubber bands. And like rubber bands, they vibrated.

Veneziano's formula thus was a doorway to a world in which the strongly interacting particles were all rubber bands, vibrating as they traveled, colliding with one another and exchanging energy. The various states of vibration would correspond to the various kinds of particles produced in the proton-smashing experiments.

This interpretation of Veneziano's formula was developed independently by Yoichiro Nambu at the University of Chicago, Holger Nielsen at the Niels Bohr Institute, and Leonard Susskind, now at Stanford University. Each thought he had done something fascinating, but they found there was little interest in their work. Susskind received a rejection from *Physical Review Letters* indicating that his insight was not significant enough to publish. As he later put it in an interview, "Boom! I felt like I had gotten hit over the head with a trashcan, and I was very, very deeply upset."[2]

But a few people did get it and began to work on the same interpretation. It would perhaps have been more accurate to call the ensuing set of ideas *rubber-band theory*. But as that lacked a certain dignity, it was *string theory* that was born.

As a theory of the strongly interacting particles, string theory was after a time supplanted by the standard model. But this does not mean that string theorists were wrong; in fact, the strongly interacting particles do behave a lot like strings. As discussed in chapter 4, the force between the quarks is now described most fundamentally by a gauge field, and the basic law seems to be given by quantum chromodynamics, or QCD, which is part of the standard model. But in some circumstances the result can be described as if there were rubber bands between the quarks. This is because the strong nuclear force is very unlike the electromagnetic force. While that force becomes weaker with distance, the force between two quarks approaches a constant strength as we pull the two quarks apart and then remains constant no matter how far apart they are after that. This is why we don't ever see free quarks in accelerator experiments, only particles made of bound quarks. However, when quarks are very close together, the force between them weakens. This is important. The string (or rubber-band) picture works only when the quarks are at a sufficient distance from one another.

The original string theorists lacked this essential insight. They imagined a world in which quarks were tied together by rubber bands, period — that is, they tried to make string theory a fundamental theory, not an approximation to anything deeper. When they tried to understand the strings *qua* strings, trouble emerged. The problems stemmed from two reasonable requirements they imposed on their theory: First, string theory should be consistent with Einstein's special theory of relativity — that is, it should respect the relativity of motion and the constancy of the speed of light. Second, it should be consistent with quantum theory.

After a few years' work, it was found that string theory, as a fundamental theory, could be consistent with special relativity and quantum theory only if several conditions were satisfied. First, the world had to have twenty-five dimensions of space. Second, there had to be a *tachyon* — a particle that goes faster than light. Third, there had to be particles that could not be brought to rest. We refer to these as *massless particles*, because mass is the measure of a particle's energy when it is motionless.

The world does not appear to have twenty-five dimensions of space. Why it is that the theory was not just abandoned then and there is one of the great mysteries of science. What is certain is that this reliance on extra dimensions deterred many people from taking string theory seriously before 1984. A lot rests on who was right — the people who rejected the idea of extra dimensions before 1984 or those who became convinced of their existence afterward.

The tachyons also posed a problem. They had never been seen; even worse, their presence signaled that the theory was unstable and, quite possibly, inconsistent. It was also the case that there are no strongly interacting particles with no mass, so the theory failed as a theory of the strongly interacting particles.

There was a fourth problem. String theory contained particles, but not all the particles in nature. There were no fermions — and thus no quarks. This was a huge problem for an alleged theory of the strong interactions!

Three of the four problems were addressed in a single move. In 1970, the theorist Pierre Ramond found a way to alter the equations describing a string, so that it would have fermions.[3] He found that the theory would be consistent only if it had a new symmetry. This symmetry would mix the old particles with the new ones — that is, it would mix bosons with fermions. This was how Pierre Ramond discovered supersymmetry; thus, whatever the fate of string theory, it proved to be one route to the discovery of supersymmetry, so as an incubator of new ideas, it was already fruitful.

The new supersymmetric string theory also addressed two other problems. It had no tachyons, so that major obstacle to taking strings seriously was eliminated. And there were no longer twenty-five dimensions of space, just nine. Nine is not three, but it is closer. With time added, the new supersymmetric string (or *superstring*, for short) lives in a world of ten dimensions. This is one less than eleven, which, strangely, is the maximum number of dimensions for which one can write a theory of supergravity.

At about the same time, a second way to put fermions into the string was invented by Andrei Neveu and John Schwarz. Like Ramond's, their version of the theory had no tachyons and lived in a world with nine spatial dimensions. Neveu and Schwarz also found that they could get the superstrings to interact with one another, and they got formulas that were consistent with the principles of quantum mechanics and special relativity.

So there was just one puzzle left. How could the new supersymmetric theory be a theory of the strong interactions if it contained massless particles? But in fact there do exist bosons with no mass. One is the photon. A photon never sits still and it can go only at the speed of light. So it has energy but no mass. The same is true of the graviton, the hypothetical particle associated with gravitational waves. In 1972, Neveu and another French physicist, Joël Scherk, found that the superstring had states of vibrations corresponding to gauge bosons, including the photon. This was a step in the right direction.[4]

But an even bigger step was taken two years later, by Scherk and Schwarz. They found that some of the massless particles predicted by the theory could actually be gravitons.[5] (The same idea occurred independently to a young Japanese physicist, Tamiaki Yoneya.[6])

The fact that string theory contained gauge bosons and gravitons changed everything. Scherk and Schwarz proposed immediately that string theory, rather than being a theory of the strong interactions, was instead the fundamental theory — the theory that unifies gravity with the other forces. To see how beautiful and simple this is, note how these photon-like and graviton-like particles arise from strings. Strings can be both closed and open. A closed string is a loop. An open string is a line; it has ends. The massless particles that might be photons come from vibrations of either open or closed strings. The gravitons come only from vibrations of closed strings, or loops.

The ends of an open string can be seen as charged particles. For example, one end could be a negatively charged particle, such as an electron; the other would then be its antiparticle, the positron, which is positively charged. The massless vibration of the string between them describes the photon that carries the electrical force between the particle and the antiparticle. Thus, you get particles and forces alike from the open strings, and if the theory is designed cleverly enough, it can produce all the forces and all the particles of the standard model.

If there are only open strings, there is no graviton, so it seems as though gravity is left out. But it turns out that you *must* include the closed strings. The reason is that nature produces collisions between particles and antiparticles. They annihilate, creating a photon. From the string point of view, this is described by the two ends of the string coming together and joining. The ends go away and you're left with a closed loop.

In fact, the particle-antiparticle annihilation and the closing of the string is necessary, if the theory is to be consistent with relativity, meaning the theory is *required* to have both open and closed strings. But this means it must include gravity. And the difference between gravity and the other forces is naturally explained, in terms of the difference between open and closed strings. For the first time, gravity plays a central role in the unification of the forces.

Isn't this beautiful? The inclusion of gravity is so compelling that a reasonable and intelligent person might easily come to believe in the theory based on this alone, whether or not there was any detailed experimental evidence for it. Especially if that person has been searching for years for a way to unify the forces, and everything else has failed.

But what gives rise to it? Is there a law that requires the ends of strings to meet and join? Herein lies one of the most beautiful features of the theory, a kind of unification of motion and forces.

In most theories, particle motion and the fundamental forces are two separate things. The law of motion tells how the particle moves in the absence of external forces. Logically there is no connection between that law and the laws that govern the forces.

In string theory, the situation is very different. The law of motion dictates the laws of the forces. This is because all forces in string theory have the same simple origin — they come from the breaking and joining of strings. Once you describe how strings move freely, all you have to do to add forces is add the possibility that a string can break into two strings. By reversing the process in time, you can

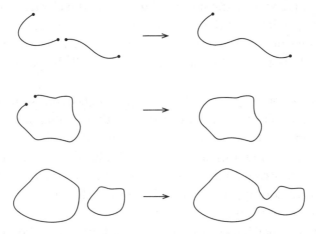

Fig 5. Top: Two open strings join at their ends. Middle: The two ends of an open string join to make a closed string. Bottom: Two closed strings join to make a single closed string.

rejoin two strings into a single string (see Fig. 5). The law for breaking and joining turns out to be strongly prescribed, to be consistent with special relativity and quantum theory. Force and motion are unified in a way that would have been impossible in a theory of particles as points.

This unification of forces and motion has a simple consequence. In a particle theory, you can freely add all kinds of forces, so there is nothing to prevent a proliferation of constants describing the workings of each force. But in string theory, there can be only two fundamental constants. One, called the *string tension*, describes how much energy is contained per unit-length of string. The other, called the *string coupling constant*, is a number denoting the probability of a string breaking into two strings, thus giving rise to a force; as it is a probability, it is a simple number, without units. All the other constants in physics must be related to these two numbers. For example, Newton's gravitational constant turns out to be related to the product of their values.

Actually, the string coupling constant is not a free constant but a physical degree of freedom. Its value depends on the solution of the theory, so rather than being a parameter of the laws, it is a parameter that labels solutions. One can say that the probability for a string

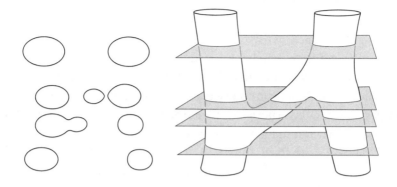

Fig. 6. The propagation and interaction of strings are determined by the same law, which is to minimize the area of the surface in spacetime. On the right, we see the surface in spacetime traced out by two closed strings, which interact by exchanging a third closed string. On the left, we see the sequence of configurations in space, which come from taking slices through the picture in spacetime on the right. First we see two closed strings, then one splits off a third closed string, which travels and then joins the second string.

to break and join is fixed not by the theory but by the string's environment — that is, by the particular multidimensional world it lives in. (This habit of constants migrating from properties of the theory to properties of the environment is an important aspect of string theory, which we will run into again in the next chapter.) On top of all of this, the law that strings satisfy is beautiful and simple. Imagine blowing a bubble. It makes a perfectly spherical shape as it expands. Or look at the bubbles next time you take a bubble bath. Their shapes are a manifestation of a simple law, which we will call the *law of bubbles.* The law states that the surface of a bubble takes up the minimal area it can, given the constraints and forces on it.

This principle turns out to apply to strings as well. As a one-dimensional string moves through time, it makes a two-dimensional surface in spacetime (see Fig. 6). This surface has a certain area, defined roughly as the product of its length and its duration in time.

The string moves so as to minimize this area. That is the whole law. It explains the motion of strings and, once strings are allowed to break and join, the existence of all the forces. It unifies all the forces we know with a description of the particles. And it is far simpler than the laws describing any of the things it unifies.

String theory achieves yet another feat of unification. In the early nineteenth century, Michael Faraday had imagined the electrical and magnetic fields in terms of *field lines* — lines running between the poles of magnets, or between positive and negative electric charges. For Faraday, these lines were real; they were what conveyed the forces between magnets or charges.

In Maxwell's theory, the field lines became secondary to the fields, but it does not have to be this way. One can imagine that the field lines are what really exist and the forces between particles are field lines stretching between them. This cannot be achieved in classical theory, but it can in quantum theory.

In a superconductor — that is, a material with little or no electrical resistance — the field lines of the magnetic field become discrete. Each line carries a certain minimal amount of magnetic flux. One can think of these field lines as a kind of atom of the magnetic field. In the early 1970s, three visionaries proposed that the same thing was true of the lines of force in QCD, which are analogous to the electric field lines of electromagnetism. This was how the Danish physicist Holger Nielsen came to be one of the inventors of string theory — he saw the strings as quantized lines of electric flux. That picture was further developed by Kenneth Wilson at Cornell, and ever since, the lines of a quantized electric field have been called Wilson lines. The third visionary was the Russian physicist Alexander Polyakov, who is perhaps our deepest thinker on the relationship between gauge theories and string theories. Polyakov gave the single most inspiring seminar I heard as a graduate student, in which he proclaimed his ambition to solve QCD exactly by re-expressing it as a theory of strings — the strings being the lines of quantized electric flux.

According to these visionaries, the primary objects in a gauge theory are the field lines. They satisfy simple laws, which dictate how they stretch between charges. The fields themselves arise only as an alternative description. This way of thinking fits naturally into string theory, because the field lines can be taken to be strings.

This suggests a kind of duality of descriptions: One can think of the field lines as the primary object and the basic laws as describing how they stretch and move, or one can think of the field as primary and the field lines just as a convenient way to describe the field. In quantum theory, either description works. This gives rise to a principle we call *the duality of strings and fields.* Either description works. Either can be taken to be fundamental.

Pierre Ramond was denied tenure at Yale in 1976, a few years after having solved several of string theory's central problems. It turns out that inventing a way to put fermions into string theory, discovering supersymmetry, and removing the tachyon — all in one blow — was not enough of an achievement to convince his colleagues that he deserved a professorship at an Ivy League institution.

John Schwarz, meanwhile, had been denied tenure at Princeton, in 1972, in spite of *his* fundamental contributions to string theory. He then moved to Caltech, where he was a research associate for the next twelve years, supported by temporary funds that had to be renewed periodically. He didn't have to teach if he didn't want to — but neither did he have tenure. He discovered the first good idea about how gravity and the other forces could be unified, but apparently Caltech remained unconvinced that he belonged on their permanent faculty.

There is no doubt that the original inventors of string theory paid heavily for their pioneering discoveries. To appreciate what kind of people these are, the reader must understand what this means in real terms. The friends you went to graduate school with are now full professors with tenure. They have good salaries, job security, they easily support families. They have high-status positions in elite universities. You have nothing. In your gut, you know that they have taken an easy road, while you have done something potentially much more significant, which has taken much more creativity and courage. They followed the herd and did what was fashionable; you discovered a whole new kind of theory. But you are still a postdoc or research associate or junior professor. You have no long-term job security and uncertain prospects. And yet you may be more active as a scientist — publishing more papers and supervising more students — than other people whose work in less risky directions has been rewarded with more security.

Now, reader, ask yourself what you would do in that situation.

John Schwarz kept working on string theory, and he continued to discover evidence that it could well be the unifying theory of physics. Although he couldn't yet prove that the theory was mathematically consistent, he was sure he was on to something.* Even as the first string theorists faced formidable obstacles, they could inspire themselves by thinking about all the puzzles that would be solved if elementary particles were vibrations of strings. It is a pretty impressive list:

1. String theory gave us an automatic unification of all the elementary particles, and it also unified the forces with one another. All come from vibrations of one fundamental kind of object.
2. String theory automatically gave us gauge fields, which are responsible for electromagnetism and the nuclear forces. These naturally arise from the vibrations of open strings.
3. String theory automatically gave us gravitons, which come from vibrations of closed strings, and any quantum theory of strings must involve closed strings. As a consequence, we got, for free, an automatic unification of gravity with the other forces.
4. A supersymmetric string theory unified the bosons and fermions, which are both just oscillations of strings, thus unifying all the forces with all the particles.

Furthermore, while supersymmetry may be true even if string theory is not, string theory provides a much more natural home for supersymmetry than do ordinary particle theories. While the supersymmetric versions of the standard model were ugly and complicated, supersymmetric string theories are deeply elegant objects.

To top it all off, string theory achieved effortlessly a natural unification of the laws of motion with the laws that govern forces.

Here then is the dream that string theory seemed to make possible. The whole standard model, with its twelve kinds of quarks and leptons and its three forces, plus gravity, could be unified, in the sense that all these phenomena arise from the vibrations of strings

*A theory is mathematically consistent when it never gives two results that contradict each other. A related requirement is that all physical qualities the theory describes involve finite numbers.

stretched in spacetime, following the simplest possible law: that the area is minimized. All the constants of the standard model can be reduced to a combination of Newton's gravitational constant and one simple number, which is the probability for a string to break into two and join. And even the second number is not fundamental but a property of the environment.

Given that string theory promised so much, it is not surprising that Schwarz and his few collaborators were convinced it must be true. As far as the problem of unification was concerned, no other theory offered so much on the basis of a single simple idea. In the face of such promise, only two questions remained: Does it work? And what is the cost?

In 1983, while I was still a postdoc at the Institute for Advanced Study in Princeton, John Schwarz was invited to give two lectures on string theory at Princeton University. I had not heard much about string theory before, and what I recall from his seminar is mostly the intense and edgy reaction of the audience, powered by equal parts interest and skepticism. Edward Witten, already a dominant figure in elementary-particle physics, interrupted often, asking a series of persistent, hard questions. I took this to be an indication of skepticism; only later would I come to see that it was an indication of his strong interest in the subject. Schwarz was confident, but there was a hint of stubbornness. I got the impression that he had spent many years trying to communicate his excitement about string theory. That talk convinced me that Schwarz was a courageous scientist, but it did not persuade me to work on string theory. For the time being, everyone I knew ignored the new theory and kept on with their various projects. Few of us realized that we were living in the last days of physics as we had always known it.

8

The First Superstring Revolution

THE FIRST SUPERSTRING revolution took place in the fall of 1984. Calling it a revolution sounds a bit pretentious, but the term is apt. Six months before, only a handful of intrepid physicists were working on string theory. They were ignored by all but a few colleagues. As John Schwarz tells it, he and a new collaborator, the English physicist Michael Green, had "published quite a few papers and in each case I was quite excited about the results. . . . [I]n each case, we felt that people would now get interested, because they could see how exciting the subject was. But there was still just no reaction."[1] Six months later, several of string theory's most vocal critics had begun working on it. In the new atmosphere, it took courage *not* to drop what you were doing and follow them.

The tipping point was a calculation carried out by Schwarz and Green providing strong evidence that string theory was a finite and consistent theory. A bit more precisely, what they finally succeeded in showing was that a certain dangerous pathology afflicting many unified theories, called an anomaly, was absent in supersymmetric string theory, at least in ten spacetime dimensions.[2] I recall that the response to that paper was both shock and jubilation: shock because some people had doubted that string theory could ever be made consistent with quantum mechanics at any level; jubilation because, by proving them wrong, Green and Schwarz had opened up the possibility that the final theory unifying physics was in our hands.

No change could have taken place quicker. As Schwarz remembers it,

> [B]efore we even finished writing it up, we got a phone call from Ed Witten saying that he had heard . . . that we had a result on canceling anomalies. And he asked if we could show him our work. So we had a draft of our manuscript at that point, and we sent it to him by FedEx. There wasn't e-mail then; it didn't exist, but FedEx did exist. So we sent it to him, and he had it the next day. And we were told that the following day everyone in Princeton University and at the Institute for Advanced Study, all the theoretical physicists, and there were a large number of them, were working on this. . . . So overnight it became a major industry [laughter], at least in Princeton — and very soon in the rest of the world. It was kind of strange, because for so many years we were publishing our results and nobody cared. Then all of a sudden everyone was extremely interested. It went from one extreme to the other: the extreme of nobody taking it seriously, to the other extreme. . . .[3]

String theory promised what no other theory had before — a quantum theory of gravity that is also a genuine unification of forces and matter. It appeared to offer, in one bold and beautiful stroke, a solution to at least three of the five great problems of theoretical physics. Thus, all of a sudden, after so many failures, we had struck gold. (Schwarz, it is amusing to note, was promptly promoted from senior research associate to full professor at Caltech.)

Thomas Kuhn, in his famous book *The Structure of Scientific Revolutions*, gave us a new way of thinking about events in the history of science that we think of as revolutions. According to Kuhn, a scientific revolution is preceded by the piling up of experimental anomalies. As a result, people begin to question the established theory. A few invent alternative theories. The revolution culminates in experimental results that favor one of the new alternatives over the old established theory.[4] It is possible to take issue with Kuhn's description of science, and I will do so in the closing section of the book. But since it describes what has happened in some cases, it serves as a useful point of comparison.

The events of 1984 did not follow Kuhn's structure. There never was an established theory addressing the problems that string theory addresses. There were no experimental anomalies; the standard model of particle physics and general relativity together sufficed to explain the results of all the experiments done until that time. Even so, how could one not call this a revolution? All of a sudden we had a good candidate for a final theory that could explain the universe and our place in it.

For four or five years after the superstring revolution of 1984, there was a lot of progress, and interest in string theory grew rapidly. It was the hottest game in town. Those who went into it dived in with ambition and pride. There were a lot of new technical tools to learn, so to work in string theory required an investment of a few months to a year, which for a theoretical physicist is a long time. Those who did it looked down on those who wouldn't, or (the suggestion was always there) couldn't. Very quickly there developed an almost cultlike atmosphere. You were either a string theorist or you were not. A few of us tried to keep a commonsense approach: Here is an interesting idea; I'll work on it some, but I'll also pursue other directions. It was hard to make that stick, because those who jumped in weren't much interested in talking with those of us who did not declare ourselves part of the new wave.

As befits a new field, immediately there were academic conferences on string theory. These had an air of triumphant celebration. There was a sense that the one true theory had been discovered. Nothing else was important or worth thinking about. Seminars devoted to string theory sprang up at many of the major universities and research institutes. At Harvard, the string theory seminar was called the Postmodern Physics seminar.

This appellation was not meant ironically. One thing that was seldom discussed in string theory seminars and conferences was how to test the theory experimentally. While a few people did worry about this, there were others who thought it wasn't necessary. The feeling was that there could be only one consistent theory that unified all of physics, and since string theory appeared to do that, *it had to be right*. No more reliance on experiment to check our theories. That was the stuff of Galileo. Mathematics now sufficed to explore the laws of nature. We had entered the period of postmodern physics.

Very quickly, physicists realized that string theory was not unique after all. Instead of a single consistent theory, we soon discovered that there were five consistent superstring theories in ten-dimensional spacetime. This gave rise to a puzzle that would not be solved for the next ten years or so. Still, it was not entirely bad news. Recall that Kaluza-Klein theory had a fatal problem: that the universes it describes are too symmetric, failing to agree with the fact that nature is not the same when viewed in a mirror. Some of the five superstring theories were able to avoid this fate and describe worlds as asymmetric as our own. And there were further developments confirming that string theory was finite (that is, that it would give only finite numbers as predictions for the result of any experiment). In the bosonic string, with no fermions, it is easy to show that there are no infinite expressions analogous to those of the theory of gravitons, but when you compute probabilities to a greater degree of precision, infinities can appear, which are related to the instability of tachyons. Since the superstring has no tachyons, this raises the possibility that the theory has no infinities.

This was easy to verify, to a low order of approximation. Beyond that, there were intuitive arguments that the theory should be finite to *every* order of approximation. I recall a prominent string theorist saying that it was so obvious that string theory was finite that he wouldn't study a proof even if there was one. But some people did endeavor to prove the finiteness of string theory past the lowest approximation. Finally, in 1992, Stanley Mandelstam, a highly respected mathematical physicist at Berkeley, published a paper that was believed to prove that superstring theories are finite to all orders of a certain approximation scheme.[5]

No wonder people were so optimistic. The promise of string theory vastly exceeded that of any unified theory so far proposed. At the same time, we could see that it still had a long way to go to fulfill all this promise. For example, consider the problem of explaining the constants of the standard model. String theory, as noted in the last chapter, has only one constant that can be adjusted by hand. *If string theory is right, the twenty constants of the standard model must be explained in terms of this one constant.* It would be marvelous beyond words if all of those constants could be computed as functions of the single constant in string theory — a triumph greater than any in the history of science. But we weren't there yet.

Beyond this was a question that, as we discussed earlier, must always be asked of unified theories. How are the apparent differences between the unified particles and forces to be explained? String theory unifies all the particles and forces, which means it must also explain to us why they are different.

So it came down, as things always do, to the details. Does it really work, or is there fine print that diminishes the miracle? If it works, how does such a simple theory actually explain so much? What must we believe about nature if string theory is true? What, if anything, do we lose in the process?

As I learned more about the theory, I began to think of the challenges it posed as very like the ones we confront when buying a new car. You go to the dealer with a list of the options you want. The dealer is overjoyed to sell you a car with those options. Several models are brought out. After a while, you realize that every car you're being shown has some options that are not on your list. You wanted antilock brakes and a really good sound system with a CD player. The cars with those also have sunroofs, fancy chrome bumpers, titanium hubcaps, eight cupholders, and custom racing stripes.

This is what is known as the package deal. It turns out that you cannot get a car with only the options you want. You have to get a package of options, which includes things you don't want or need. These extras raise the price quite a bit, but there is no choice. If you want the antilock brakes and the CD player, you have to take the whole package.

String theory, too, seems to be offered only as a package deal. You may want a simple unified theory of all the particles and forces, but what you get includes a few extra features, at least two of which are nonnegotiable.

The first is supersymmetry. There were string theories without supersymmetry, but they were all known to be unstable, because of the presence of those pesky tachyons. Supersymmetry appeared to eliminate the tachyons, but there was a catch. The supersymmetric string theory could be consistent only if the universe has nine dimensions of space. There was no option for a theory that works in a three-dimensional space. If you wanted the other features, you had to take the option with six extra dimensions. Much followed from this. If the theory was not to be ruled out right away, there had to be a way to hide the extra dimensions. There appeared to be no choice but to curl them up so that they were too small to be perceived. Thus we were forced to revive the main ideas of the old unified-field theories.

This gave rise to great opportunities, and great problems. As we saw, earlier attempts to use higher dimensions to unify physics had failed, because there were too many solutions; the introduction of the higher dimensions led to a huge problem of nonuniqueness. It also led to problems of instabilities, because there are processes by which the geometry of the extra dimensions unravels and becomes large and other processes whereby it collapses to a singularity. If string theory was to succeed, it would have to solve these problems.

String theorists soon realized that the problem of nonuniqueness was a fundamental feature of string theory. There were now six extra dimensions to curl up, and there were many ways to do it. Most involved a complicated six-dimensional space, and each gave rise to a different version of string theory. Because string theory is a background-dependent theory, what we understood about it at a technical level was that it gave us a description of strings moving in fixed-background geometries. By choosing different-background geometries, we got technically different theories. They came from the same idea, and the same law was applied in each case. But strictly speaking, each was a different theory.

This is not just splitting hairs. The physical predictions given by all these different theories were different, too. Most of the six-dimensional spaces were described by a list of constants, which were freely specifiable. These denoted different features of the geometry, such as the volumes of the extra dimensions. A typical string theory might have hundreds of these constants. These constants are part of the description of how strings propagate and interact with each other.

Think of an object with a two-dimensional surface, like a sphere. Because it is perfectly spherical, it is described by only one parame-

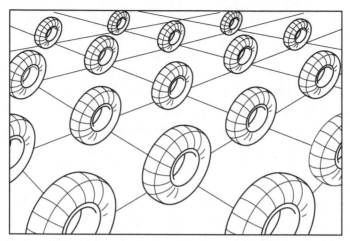

Fig. 7. The hidden dimensions can have different topologies.
In this example, there are two hidden dimensions, which
have the topology of a doughnut, or torus.

ter, its circumference. But now imagine a more complicated surface, like a doughnut (see Fig. 7). This surface is described by two numbers. There are two circles that go around the doughnut in two different ways, and they can have different circumferences.

We can imagine more complicated surfaces, with many holes. These take more numbers to describe. But no one (at least, no one I know) can directly visualize a six-dimensional space.

However, we do have tools to describe them, which employ analogs of the holes that can occur in a doughnut and other two-dimensional surfaces. Rather than wrapping a string around a hole, we wrap a higher-dimensional space around it. In each case, the space that is wrapped will have a volume, and that will become a constant describing the geometry. When we work out how the strings move in the extra dimensions, all these extra constants come in. So there is no longer just one constant; there are a large number of constants.

This is how string theory resolves the basic dilemma facing attempts to unify physics. Even if everything comes from a simple principle, you have to explain how the variety of particles and forces arises. In the simplest possibility, where space has nine dimensions, string theory is very simple; all the particles of the same kind are identical. But when the strings are allowed to move in the complicated geometry of the six extra dimensions, there arise lots of different kinds of particles, associated with different ways to move and vibrate in each of the extra dimensions.

So we get a natural explanation for the apparent differences among the particles, something a good unified theory must do. But there is a cost, which is that the theory turns out to be far from unique. What is happening is a trading of constants: The constants that denote the masses of the particles and the strengths of the forces are being traded for constants that denote the geometry of the extra six dimensions. It is then less surprising to find constants that would explain the standard model.

Even so, this scheme might have been compelling if it had led to unique predictions for the constants of the standard model. If by translating the standard model's constants into constants denoting the geometry of the extra dimensions, we had found out something new about the standard model's constants, and if these findings had agreed with nature, that would have constituted strong evidence that string theory must be true.

But this is not what happened. The constants that could be freely varied in the standard model were translated into geometries that could be freely varied in string theory. Nothing was constrained or reduced. And because there were a huge number of choices for the geometry of the extra dimensions, the number of free constants went up, not down.

Furthermore, the standard model was not completely reproduced. It is true that we can derive its general features, such as the existence of fermions and gauge fields. But the exact combinations seen in nature did not come out of the equations.

From here it got worse. All the string theories predicted extra particles — particles not seen in nature. Along with them came extra forces. Some of these extra forces came from variations in the geometry of the extra dimensions. Think of a sphere attached to every point in space, as in Fig. 8. The radius of the sphere can change as we move around in space.

So the radius of each sphere can be seen as a property of the point to which it is attached. That is, it is something like a field. Just like the electromagnetic field, such fields propagate in space and time, and this gives rise to extra forces. This is clever, but there was a danger that these extra forces would disagree with observation.

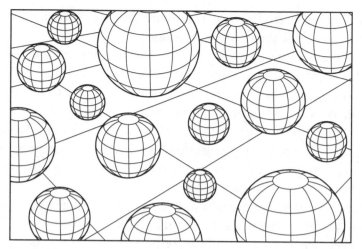

Fig. 8. The geometry of the hidden dimensions can vary in space and time. In this example, the radii of the spheres vary.

We have been speaking of generalities, but there is one world. If string theory was to succeed, it had to not just model possible worlds but also explain *our* world. A key question, then, was: *Is there a way to curl up the extra six dimensions so that the standard model of particle physics is completely reproduced?*

One way was to have a world with supersymmetry. While string theory had supersymmetry, how exactly that symmetry was manifested in our three-dimensional world turned out to be dependent on the geometry of the extra dimensions. One could arrange them so that the supersymmetry seemed to be broken in our world. Or it could be the case that there was much more supersymmetry than could be accommodated in a realistic theory.

So there arose an interesting problem: Could the geometry of the extra six dimensions be chosen so as to achieve exactly the right amount of supersymmetry? Could we arrange it so that our three-dimensional world has a version of particle physics described by the supersymmetric versions of the standard model?

This question was solved in 1985 in a very important paper, written by a quartet of string theorists: Philip Candelas, Gary Horowitz, Andrew Strominger, and Edward Witten.[6] They were lucky, because two mathematicians, Eugenio Calabi and Shing-tung Yau, had already solved a mathematical problem that gave the answer. They had discovered and studied a particularly beautiful form of six-dimensional geometry that we now call Calabi-Yau spaces. The four string theorists were able to show that the conditions needed for string theory to reproduce a version of the supersymmetric standard model were the same as the conditions that defined a Calabi-Yau space. They then proposed that nature is described by a string theory in which the extra six dimensions are chosen to be a Calabi-Yau space. This cut down on the possibilities and gave the theory more structure. For example, they showed explicitly how you could trade constants in the standard model, such as those that determine the masses of the different particles, for constants describing the geometry of a Calabi-Yau space.

This was great progress. But there was an equally great problem. Had there been only one Calabi-Yau space, with fixed constants, we would have had the unique unified theory we yearned for. Unfortunately, there turned out to be many Calabi-Yau spaces. No one knew how many, but Yau himself was quoted as saying there were at least a hundred thousand. Each of these spaces gave rise to a different version of particle physics. And each space came with a list of free constants governing its size and shape. So there was no uniqueness, no new predictions, and nothing was explained.

In addition, the theories involving Calabi-Yau spaces had lots of extra forces. It turns out that as long as string theory is supersymmetric, many of these forces will have infinite range. This was unfortunate, because there are strict experimental limits on the existence of any infinite-range force besides gravity and electromagnetism.

There remained other problems. The constants that give the geometry of the extra dimensions can vary continuously. This could give rise to instabilities, as in the old Kaluza-Klein theories. Unless there is some mysterious mechanism that freezes the geometry of the extra dimensions, these instabilities lead to catastrophe, such as singularities coming from the collapse of the extra dimensions.

On top of this, even if our world was described by one of the Calabi-Yau geometries, there was no explanation for how it got that way. String theory came in many other versions besides the Calabi-Yau spaces. There were versions of the theory in which the number of curled-up dimensions varied from none all the way up to nine.

Those geometries that had dimensions that weren't curled up were called *flat*; they defined worlds that large beings like us would experience. (In investigating the implications for particle physics, we could ignore gravity and cosmology, in which case the non-curled-up dimensions had a geometry described by the special theory of relativity.)

A hundred thousand Calabi-Yau manifolds were only the tip of the iceberg. In 1986, Andrew Strominger discovered a way to construct a vast number of additional supersymmetric string theories. It will be useful to keep in mind what he wrote in the conclusion of his paper describing that construction:

> [T]he class of supersymmetric superstring compactifications has been enormously enlarged. . . . [I]t does not seem likely that [these] solutions . . . can be classified in the foreseeable future. As the constraints on [these] solutions are relatively weak, it does seem likely that a number of phenomenologically acceptable . . . ones can be found. . . . While this is quite reassuring, in some sense life has been made too easy. *All predictive power seems to have been lost.*
>
> All of this points to the overwhelming need to find a dynamical principle for determining [which theory describes nature], which now appears more imperative than ever.[7] (Italics mine.)

Thus, by taking on the strategy of the older higher-dimensional theories, string theory took on their problems as well. There were lots of solutions, and a few of them gave rise to a description of something roughly like the real world, but most didn't. There were lots of instabilities, which manifested themselves in lots of extra forces and particles.

This was bound to create controversy, and it did. Few could disagree that the list of good features was long and impressive. It really did seem that the idea of particles as vibrations of strings was the missing link that could work powerfully to resolve many open problems. But the price was high. The extra features we were forced to buy took away some of the beauty of the original proposal — at least, for a few of us. Others found the geometry of the extra dimensions the most beautiful thing about the theory. No wonder theorists came down strongly on either side.

Those who believed tended to believe in the whole package. I knew many physicists who were sure that supersymmetry and the extra dimensions were there, waiting to be discovered. I knew as many who jumped ship at that point, because it meant accepting too much that had no foundation in experiment.

Among the detractors was Richard Feynman, who explained his reluctance to go along with the excitement as follows:

> I don't like that they're not calculating anything. I don't like that they don't check their ideas. I don't like that for anything that disagrees with an experiment, they cook up an explanation — a fix-up to say "Well, it still might be true." For example, the theory requires ten dimensions. Well, maybe there's a way of wrapping up six of the dimensions. Yes, that's possible mathematically, but why not seven? When they write their equation, the equation should decide how many of these things get wrapped up, not the desire to agree with experiment. In other words, there's no reason whatsoever in superstring theory that it isn't eight of the ten dimensions that get wrapped up and that the result is only two dimensions, which would be completely in disagreement with experience. So the fact that it might disagree with experience is very tenuous, it doesn't produce anything; it has to be excused most of the time. It doesn't look right.[8]

These sentiments were shared by many of the older generation of particle physicists, who knew that the success of particle theory had always required a continual interaction with experimental physics. Another dissenter was Sheldon Glashow, Nobel Prize winner for his work on the standard model:

> But superstring physicists have not yet shown that their theory really works. They cannot demonstrate that the standard theory is a logical outcome of string theory. They cannot even be sure that their formalism includes a description of such things as protons and electrons. And they have not yet made even one teeny-tiny experimental prediction. Worst of all, superstring theory does not follow as a logical consequence of some appealing set of hypotheses about nature. Why, you may ask, do the string theorists insist that space is nine-dimensional? Simply because string theory doesn't make sense in any other kind of space. . . .[9]

Beyond the controversy, however, there was a clear need to understand the theory better. A theory that came in so many different versions did not seem like a single theory. If anything, the different theories seemed like different solutions to some other, as yet unknown theory.

We are used to the idea that one theory has many different solutions. Newton's laws describe how a particle moves in response to forces. Suppose we fix the forces — for example, we want to describe a ball being thrown in Earth's gravitational field. Newton's equations have an infinite number of solutions, corresponding to the infinite number of trajectories the ball can take: It can be thrown higher or lower, faster or slower. Each way of throwing the ball gives rise to a different trajectory, each of which is a solution of Newton's equations.

General relativity also has an infinite number of different solutions, each of which is a spacetime — that is, a possible history of the universe. Since the geometry of spacetime is a dynamical entity, it can exist in an infinity of different configurations and evolve into an infinity of different universes.

Each of the backgrounds on which a string theory is defined is a solution of Einstein's equation or some generalization of it. Thus, it began to occur to people that the growing catalog of string theories meant that we weren't actually studying a fundamental theory. What we were doing, perhaps, was studying the *solutions* to some deeper, still unknown theory. We might call this a *meta-theory*, because each of its solutions is a theory. This meta-theory is the real fundamental law. Each solution of it will give rise to a string theory.

Thus, it would be more compelling if we could think not of an infinity of string theories but of an infinity of solutions arising from one fundamental theory.

Recall that each of the many string theories is a background-dependent theory that describes strings moving in a particular background spacetime. Since the various approximate string theories live on different spacetime backgrounds, the theory that unifies them *must not live on any spacetime background*. What is needed to unify them is a single, background-independent theory. The way to do this was thus clear: Invent a meta-theory that would itself be background-independent, then derive all the background-dependent string theories from this single meta-theory.

So we had two reasons to look for a background-independent quantum theory of gravity. We already knew that we had to incorporate the dynamical character of geometry given in Einstein's general theory of relativity. Now we needed it to unify all the different string theories. Doing this would require a new idea but, at least for the time being, it remained out of reach.

One thing that the meta-theory was expected to do was help select which version of string theory was realized physically. Since it was widely believed that string theory was the unique unified theory, many theorists expected that most of the large number of variants would be unstable and that the one truly stable theory would uniquely explain the standard-model constants.

Sometime in the late 1980s, it occurred to me that there was another possibility. Perhaps all string theories were equally valid. This would imply a complete revision of our expectations about physics, in that it would make all the properties of the elementary particles contingent — determined not by fundamental law but by one of an infinite number of solutions to the fundamental theory. There were already indications that this contingency could happen in theories with spontaneous symmetry breaking, but the many versions of string theory opened up the possibility that it was true of essentially all the properties of the elementary particles and forces.

This would mean that the properties of the elementary particles were environmental and could change in time. If so, it would mean that physics would be more like biology, in that the properties of the elementary particles would depend on the history of our universe. String theory would not be one theory, it would be a landscape of theories — analogous to the fitness landscapes that evolutionary biologists study. There might even be a process like natural selection that would select which version applied to our universe. (These thoughts would lead to a 1992 paper titled "Did the Universe Evolve?"[10] and a 1997 book called *The Life of the Cosmos.* Our story later turns on these ideas.)

Whenever I discussed this evolutionary principle with string theorists, they would say, "Don't worry, there will be a unique version of string theory, selected by a so far unknown principle. When we find it, this principle will correctly explain all the parameters of the standard model and lead to unique predictions for upcoming experiments."

In any event, progress on string theory slowed, and by the early

1990s string theorists were discouraged. There was no complete formulation of string theory. All we had was a list of hundreds of thousands of distinct theories, each with many free constants. We had no precise idea which of the many versions of the theory corresponded to reality. And while there had been great technical progress, no smoking gun had emerged that would tell us whether string theory was right or wrong. Worst of all, there was not a single prediction made that might be confirmed or falsified by a doable experiment.

There were other reasons for string theorists to be discouraged as well. The late 1980s had been good for the field. Just after the revolution of 1984, the inventors of string theory, like John Schwarz, had had many tempting offers from the best universities. For a few years, young string theorists had moved ahead. But by the early 1990s, this had fallen off, and talented people were again going without job offers.

Some people, young and old, left the field at this point. Luckily, working on string theory had proved to be good intellectual training, and some former string theorists are now flourishing in other areas, such as solid-state physics, biology, neuroscience, computers, and banking.

But others stayed the course. Despite all the reasons for discouragement, many string theorists could not let go of the idea that string theory constituted the future of physics. If there were problems, well, no other approach to unifying the elementary particles was succeeding either. There were a few people working on quantum gravity, but most string theorists remained blissfully unaware of them. For many of them, string theory was simply the only game in town. Even if it was a harder road than they had hoped it would be, no other theory promised to unify all the particles and forces and solve quantum gravity, all within a finite and consistent framework.

The unfortunate result was that the split between believers and skeptics deepened. Each side became more entrenched, and each seemed to have good justification for its position. And it would have stayed like this for a long time, had certain dramatic developments not occurred that radically altered our appreciation of string theory.

9

Revolution
Number Two

STRING THEORY INITIALLY proposed to unify all the particles and forces in nature. But as it was studied in the decade following the 1984 revolution, something unexpected happened. The alleged unified theory fractured into many different theories: the five consistent superstring theories in ten-dimensional spacetime, plus millions of variants in the cases where some dimensions were wrapped up. As time went on, it became clear that string theory itself was in need of unification.

The second superstring revolution, which burst on the scene in 1995, gave us just that. The birth of the revolution is often taken to be a talk that Edward Witten gave that March at a string theory conference in Los Angeles, where he proposed a unifying idea. He did not actually present a new unified superstring theory; he simply proposed that it existed and that it would have certain features. Witten's proposal was based on a series of recent discoveries that had uncovered new facets of string theory and greatly increased our understanding of it. These had further unified string theory with gauge theories and general relativity by exposing additional deep commonalities and relationships among them. These advances, several of which were unprecedented in the history of modern theoretical physics, had eventually won over many skeptics, including me. At first, it had appeared that the five consistent superstring theories described different worlds, but in the mid 1990s we began to understand that they were not as different as they seemed.

When two different ways of looking at the same phenomenon arise, we refer to this as a *duality*. Ask the members of a couple to tell you, separately, the story of their relationship. They will not be the same stories, but each important event in one will correspond to an important event in the other. If you talk to them long enough, you will be able to predict how the two stories relate and differ. For example, a husband's perception of his wife's assertiveness might map to the wife's perception of an instance of her husband's passivity. One can say that the two descriptions are *dual* to each other.

String theorists, in their efforts to relate the five theories to one another, began to speak of several kinds of dualities. Some dualities are *exact:* that is, the two theories are not really different but are simply two ways of describing the same phenomenon. Other dualities are *approximate.* In these cases, the two theories really are different, but there are phenomena in one that are similar to phenomena in the other, leading to approximations in which certain features of one theory can be understood by studying the other.

The simplest duality that holds among the five superstring theories is called *T-duality.* "T" stands for "topological," because this duality has to do with the topology of space. It occurs when one of the compactified dimensions is a circle. In this case, a string can wind around the circle; in fact, it can wind a number of times (see Fig. 9). The number of times the string wraps around the circle is called the *winding number.*

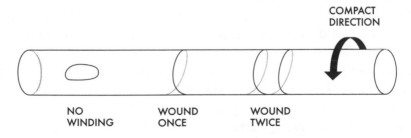

COMPACT
DIRECTION

NO
WINDING

WOUND
ONCE

WOUND
TWICE

Fig. 9. Strings can wind around a hidden dimension. In this case, space is one-dimensional and the hidden dimension is a small circle. Pictured are strings that wind around the circle zero, one, and two times.

Another number measures how a string is vibrating. A string has overtones, just like a piano string or a guitar string, and natural numbers denote the various levels of vibration. T-duality is a relationship between two string theories both of which wrap around a circle. The radii of the two circles differ but are related to each other; one is equal to the inverse of the other (in units of string length). In such cases, the winding states of the first string theory behave exactly the same as the levels of vibration of the second string theory. This kind of duality turns out to exist between certain pairs of the five string theories. They appear to be different theories to start with, but when you wrap their strings around circles, they become the same theory.

There is a second kind of duality that is also conjectured to be exact, although this has not yet been proved. Recall from chapter 7 that there is a constant in each string theory that determines how probable it is that strings will break and join. This is the string coupling constant, conventionally denoted by the letter g. When g is small, the probability for strings to break and join is small, so we say the interactions are weak. When g is large, they break and join all the time, so we say the interactions are strong.

Now, it can happen that two theories are related in the following way: Each theory has a coupling g. But when the g of the first theory is equal to $1/g$ of the second theory, the two theories appear to behave identically. This is called S- (for *strong-weak*) *duality*. If g is small, meaning the strings interact weakly, $1/g$ is big, so the strings in the second theory interact strongly.

How can these two theories behave identically if their coupling constants are different? Can't we tell if the probability for strings to join and break is large or small? We can, if we know what the strings are. But what is believed to happen in cases of S-duality is that these two theories have more strings than they are supposed to.

This proliferation of strings is an example of the familiar but rarely understood phenomenon of *emergence*, a term that describes the arising of new properties in large and complex systems. We may know the laws that the elementary particles satisfy, but when many particles are bound together, all kinds of new phenomena become apparent. A bunch of protons, neutrons, and electrons may combine to produce a metal; others, of equal number, may combine to produce a living cell. Both the metal and the living cell are just collections of protons, neutrons, and electrons. How, then, do we describe what makes a metal a metal and a bacterium a bacterium? The properties that distinguish them are called *emergent properties*.

Here's an example: Perhaps the simplest thing a metal can do is vibrate; if you hit one end of a metal bar, a sound wave will travel through it. The frequency at which the metal vibrates is an emergent property, as is the speed that sound travels in the metal. Recall the wave/particle duality of quantum mechanics, which asserts that there is a wave associated with every particle. The reverse is also true: There is a particle associated with every wave, including a particle associated with the sound wave traveling through the metal. It is called a *phonon*.

A phonon is not an elementary particle. It is certainly not one of the particles that make up the metal, for it exists only by virtue of the collective motion of huge numbers of the particles that do make up the metal. But a phonon is a particle just the same. It has all the properties of a particle. It has mass, it has momentum, it carries energy. It behaves precisely the way quantum mechanics says a particle should behave. We say that a phonon is an *emergent particle*.

Things like this are believed to happen to strings as well. When the interactions are strong, there are many, many strings breaking and joining, and it becomes difficult to follow what happens to each individual string. We then look for some simple emergent properties of large collections of strings — properties that we can use to understand what is going on. Now comes something really fun. Just as the vibrations of a whole bunch of particles can behave like a simple particle — a phonon — a new string can emerge out of the collective motion of large numbers of strings. We can call this an *emergent string*.

The behavior of these emergent strings is the exact opposite of that of ordinary strings — let's call the latter the *fundamental strings*. The more the fundamental strings interact, the less the emergent strings do. To put this a bit more precisely: If the probability for two fundamental strings to interact is proportional to the string coupling constant g, then in some cases the probability for the emergent strings to interact is proportional to $1/g$.

How do you tell the fundamental strings from the emergent

strings? It turns out that you can't — at least, in some cases. In fact, you can turn the picture around and see the emergent strings as fundamental. This is the fantastic trick of strong-weak duality. It is as if we could look at a metal and see the phonons — the quantum sound waves — as fundamental and all the protons, neutrons, and electrons making up the metal as emergent particles made up of phonons.

Like T-duality, this kind of strong-weak duality turned out to relate certain pairs of the five superstring theories. The only question was whether this relationship applied just to some states of the theories or was deeper. This was an issue, because to show the relationship at all, you had to study special states of the paired theories — states constrained by a certain symmetry. Otherwise you didn't have enough control of the calculations to get good results.

There were, then, two possible paths for theorists. The optimists — and in those days most string theorists were optimists — went beyond what could be shown, to a conjecture that the relationship between the special symmetric states they could examine in the paired theories extended to *all five* of the theories. That is, they posited that even without the special symmetries there were always emergent strings and that they always behaved just like the fundamental strings of another theory. This implied that S-duality does not just relate some aspects of the theories but demonstrates their complete equivalence.

On the other hand, the few pessimists worried that perhaps the five theories really were different from one another. They thought it pretty wonderful that there were even a few cases in which emergent strings of one theory behaved like fundamental strings of another, but they realized that such a thing might be true even if the theories were all different.

A lot rested (and continues to rest) on whether the optimists or the pessimists were right. If the optimists turn out to be right, then all five of the original superstring theories really are just different ways of describing a single theory. If the pessimists are right, then they really all are different theories, and therefore there is no uniqueness, no fundamental theory. As long as we do not know whether strong-weak duality is approximate or exact, we do not know whether string theory is unique or not.

One piece of evidence in favor of the optimistic view was that similar dualities were known to exist in theories that were simpler and better understood than string theories. One example is a version of Yang-Mills theory called $N = 4$ super-Yang-Mills theory, which has as much supersymmetry as possible. For short, we'll call it the *maximally super theory*. There is good evidence that this theory has a version of S-duality. It works roughly like this. The theory has in it a number of electrically charged particles. It also has some emergent particles that carry magnetic charges. Now, normally there are no magnetic charges, there are only magnetic poles. Every magnet has two, and we refer to them as north and south. But in special situations there may be emergent magnetic poles that move independently of each other — they are known as *monopoles*. What happens in the maximally super theory is that there is a symmetry within which electric charges and magnetic monopoles trade places. When that happens, if you change the value of the electric charge to 1 divided by the original value, you don't change anything in the physics described by the theory. The maximally super theory is a remarkable theory, and it was to play a central role in the second superstring revolution, as we will see shortly. But now that we understand a little about different kinds of dualities, I can explain the conjecture that Witten discussed in his celebrated talk in Los Angeles.

As I mentioned, the key idea in Witten's talk was that the five consistent superstring theories were all actually the same theory. But what *was* this single theory? Witten didn't tell us, but he did describe a dramatic conjecture about it, which was that the theory unifying the five superstring theories would require one more dimension, so that space now had ten dimensions and spacetime eleven.[1]

This particular conjecture had been first made by two British physicists, Christopher Hull and Paul Townsend, a year earlier.[2] Witten had found a great deal of evidence for the conjecture, based on dualities that had been found not just between the five theories but between string theories and theories in eleven dimensions.

Why should a unification of string theories have one more dimension? A property of an extra dimension — the radius of the extra circle in Kaluza-Klein theory — can be interpreted as a field varying over the other dimensions. Witten used this analogy to suggest that a certain field in string theory was actually the radius of a circle extending in the eleventh dimension.

How did this introduction of yet one more spatial dimension help? After all, there wasn't a consistent supersymmetric string theory in eleven spacetime dimensions. But there *was* a supersymmetric gravity theory in eleven spacetime dimensions. This, you may recall from chapter 7, is the highest-dimensional of all the supergravity theories, a veritable Mount Everest of supergravity. So Witten conjectured that the eleven-dimensional world whose existence the extra field pointed to could be described — in the absence of quantum theory — by eleven-dimensional supergravity.

Moreover, although there isn't a string theory in eleven dimensions, there *is* a theory of two-dimensional surfaces moving in an eleven-dimensional spacetime. This theory is quite beautiful, at least at the classical level. It was invented in the early 1980s and is called, imaginatively, the *eleven-dimensional supermembrane theory*.

The supermembrane theory had been ignored by most string theorists until Witten, and for good reason. It is not known whether the theory can be made consistent with quantum mechanics. Some people had tried to combine it with quantum theory and failed. When the first superstring revolution took off in 1984, based on magical properties of theories in ten dimensions, these eleven-dimensional theories were given up by most theorists.

But now, following Witten, string theorists proposed to revive the membrane theory in eleven dimensions. They did so because they noticed several amazing facts. First, if you take one of the eleven dimensions to be a circle, then you can wrap one dimension of a membrane around that circle (see Fig. 10). This leaves the other dimension of the membrane free to move in the remaining nine dimensions of space. This is a one-dimensional object moving in a nine-dimensional space. It looks just like a string!

Witten found that you could get all five of the consistent superstring theories by wrapping one dimension of a membrane in different ways around the circle; moreover, you got those five theories and no others.

This is not all. Recall that when a string is wrapped around a circle, there are transformations called T-dualities. As opposed to

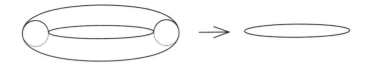

Fig. 10. On the left, we have a two-dimensional membrane, which we can imagine is wrapped around a hidden dimension, which is a small circle. Seen from far enough away (right), it looks like a string wrapped around the large dimension.

other kinds of dualities, these are known to be exact. We also find such dual transformations when one dimension of a membrane is wrapped around a circle. If we interpret these transformations in terms of the string theories we get from wrapping the membrane, they turn out to be exactly the strong-weak dualities that connect those string theories. Those particular dualities, you'll recall, had been conjectured but not proved, outside of special cases. They were now understood to come from transformations of the eleven-dimensional theory. This is so pretty that it's hard not to believe in the existence of the eleven-dimensional unifying theory. The only problem left open was to discover it.

Later that year, Witten gave the so-far-undefined theory a name. The act of naming it was brilliant: He called it simply *M-theory*. He didn't want to say what "M" stood for, because the theory did not yet exist. We were invited to fill in the rest of the name by inventing the theory itself.

Witten's talk raised many questions. If he was right, there was a lot to discover. One person listening was Joseph Polchinski, a string theorist working in Santa Barbara. As he tells it, "After Ed's talk, I made a list of twenty homework problems for myself, to understand it better."[3] The homework led him to a discovery that would be key in the second superstring revolution — that string theory is not just a theory of strings. Other objects live in the ten-dimensional spacetime.

People who don't know much about aquariums think they are only about fish. But aquarium enthusiasts know that the fish are only what first attract your eye. A healthy aquarium is all about the plant life. If you try to stock an aquarium with just fish, it won't go well. You will soon have a piscine morgue. It turns out that during the first superstring revolution, from 1984 to 1995, we were like amateurs trying to make aquariums with just fish. We missed most of what was necessary to make the system work until Polchinski discovered the missing essentials.

In the fall of 1995, Polchinski showed that a string theory, to be consistent, must include not only strings but surfaces of higher dimensions moving in the background space.[3] These surfaces are also dynamical objects. Just like strings, they are free to move in space. If a string, which is a one-dimensional object, can be fundamental, why can't a two-dimensional surface be fundamental? In higher dimensions, where there is a lot of room, why not a three-, four-, or even five-dimensional surface? Polchinski found that the dualities between string theories would not work out consistently unless there were higher-dimensional objects in the theory. He called them *D-branes*. (The term "brane" comes from "membrane," which is a two-dimensional surface; the "D" refers to a technicality I won't try to explain here.) Branes play a special role in the life of the strings: They are places where open strings can end. Normally, the ends of open strings travel freely through space, but sometimes the ends of a string can be constrained to live on the surface of a brane (see Fig. 11). This is because branes can carry electric and magnetic charges.

From the point of view of the strings, the branes are additional features of the background geometry. Their existence enriches string theory by greatly increasing the number of possible background geometries where a string could live. Besides wrapping the extra dimensions in some complicated geometry, you can wrap the branes around loops and surfaces in that geometry. You can have as many branes as you like, and they can wrap around the compactified dimensions an arbitrary number of times. In this way, you can make an infinite number of possible backgrounds for string theories. This scheme of Polchinski's was to have enormous consequences.

The branes also deepen our understanding of the relationship between gauge theories and string theories. They do this by allowing a new way for symmetries to arise in string theory, a result of piling several branes one on top of another. As I just mentioned, open strings can end on the branes. But if several branes are in the same

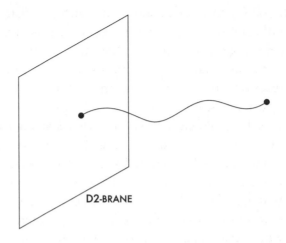

D2-BRANE

Fig. 11. A two-dimensional brane, on which an open string ends.

place, it doesn't matter which of them a string ends on. This means that there is a kind of symmetry at work, and symmetries, as described in chapter 4, give rise to gauge theories. Consequently, we find a new connection between string theory and gauge theories.

Branes also opened up a whole new way of thinking about how our three-dimensional world might relate to the extra spatial dimensions of string theory. Some of the branes that Polchinski discovered are three-dimensional. By piling up three-dimensional branes, you get a three-dimensional world with whatever symmetries you like, floating in a higher-dimensional world. Could our three-dimensional universe be such a surface in a higher-dimensional world? This is a big idea, and it makes a possible connection to a field of research called *brane worlds*, in which our universe is seen as a surface floating in a higher-dimensional universe.

Branes did all this, but they did even more. They made it possible to describe some special black holes within string theory. This discovery, by Andrew Strominger and Cumrun Vafa in 1996, was perhaps the greatest accomplishment of the second superstring revolution.

The relationship of branes to black holes is indirect but powerful. Here is how it goes: You begin by turning off the gravitational force (you do this by setting the string coupling constant to zero). It may seem strange to describe black holes, which are nothing but gravity, in this way, but watch what happens. With gravity turned off, we can consider geometries in which many branes are wrapped around the extra dimensions. We now draw on the fact that the branes carry electric and magnetic charges. It turns out that there's a limit to how much charge a brane can have, which is related to the brane's mass. The configurations with the highest possible charge are very special and are called *extremal*. These comprise one of the special situations we talked about before, where there are extra symmetries that allow us to do more precise calculations. In particular, such situations are characterized by having several different supersymmetries that relate fermions and bosons.

There is also a maximal amount of electric or magnetic charge that a black hole can have and still be stable. These are called *extremal black holes*, and they had been studied for many years by specialists in general relativity. If you study particles moving on these backgrounds, you also find several different supersymmetries.

Surprisingly, despite the fact that the gravitational force has been turned off, the extremal brane systems turn out to share some properties with extremal black holes. In particular, the thermodynamic properties of the two systems are identical. Thus, by studying the thermodynamics of extremal branes wrapped around the extra dimensions, we can reproduce the thermodynamic properties of extremal black holes.

One of the challenges of black-hole physics has been to explain Jacob Bekenstein's and Stephen Hawking's discoveries that black holes have entropy and temperature (see chapter 6). The new idea from string theory is that — at least, in the case of extremal black holes — you can make progress by studying the analogous system of extremal branes wrapped around the extra dimensions. In fact, many properties of the two systems match exactly. This almost miraculous coincidence occurs because in both cases there are several different supersymmetry transformations relating fermions and bosons. These turn out to allow construction of a powerful mathematical analogy that forces the thermodynamics of the two systems to be identical.

But this was not the whole story. You could also study black holes that were *almost* extremal, in that they had slightly less charge than the maximal amount possible. On the brane side, you could also study collections of branes that had slightly less than the maximal charge. Does the correspondence between branes and black holes still hold? The answer is yes, and precisely so. As long as you stay very close to the extremal cases, the properties of the two systems match closely. This is a much stricter test of the correspondence. On each side, there are complicated and precise relationships between temperature and other quantities such as energy, entropy, and the charges. The two cases agree very well.

In 1996, I heard a young Argentinian postdoc named Juan Maldacena lecture about these results at a conference in Trieste, where I used to spend time during the summer. I was floored. The precision with which the behavior of the branes matched the physics of black holes immediately convinced me to set aside time to work on string theory again. I took Maldacena out to dinner at a pizzeria overlooking the Adriatic, and I found him to be one of the smartest and most perceptive young string theorists I had ever encountered. One thing we discussed that night over wine and pizza was whether the systems of branes might be more than just models of black holes. Did they provide a genuine explanation of the entropy and temperature of black holes?

We could not answer that question, and it has remained open. The answer depends on how significant these results are. Here we encounter the situation I described in other cases where extra symmetry led to very powerful findings. There are, again, two points of view. The pessimistic point of view holds that the relationship between the two systems is probably an accidental result of the fact that both have a lot of extra symmetry. To a pessimist, the fact that the calculations are beautiful does not imply that they lead to general insights about black holes. On the contrary, the pessimist worries that the calculations are beautiful because they depend on very special conditions that do not extend to typical black holes.

The optimist, however, believes that all black holes can be understood using the same ideas, and that the extra symmetries present in special cases simply allow us to calculate more precisely. As with strong-weak duality, we still don't know enough to decide whether the optimists or the pessimists are right. In this case, there is an added worry, which is that the piles of branes are not black holes, because the gravitational force has been turned off. It is conjectured that they would become black holes if the gravitational force were slowly turned on. In fact, this can be imagined to happen in string theory, because the strength of the gravitational force is proportional to a field that can vary in space and time. But the problem is that such a process, where the gravitational field changes in time, has always been hard for string theory to describe concretely.

As wonderful as his work in black holes was, Maldacena was only getting started. In the fall of 1997, he released an astounding paper in which he proposed a new kind of duality.[4] The dualities we have mentioned so far are between theories of the same kind, living in a spacetime of the same number of dimensions. Maldacena's revolutionary idea was that a string theory could have a dual description in terms of a gauge theory. This is astounding because a string theory is a theory of gravity, whereas a gauge theory lives in a world without gravity, on a fixed-background spacetime. Moreover, the world described by the string theory has more dimensions than the gauge theory that represents it.

One way to understand Maldacena's proposal is to recall the idea we discussed in chapter 7, in which a string theory can arise from studying the lines of flux of the electric field. Here, the electric field's lines of flux become the basic objects of the theory. Being one-dimensional, they look like strings. You can say that the lines become emergent strings. In most cases, emergent strings that arise from gauge theories do not behave like the kinds of strings that string theorists talk about. In particular, they do not appear to have anything to do with gravity, and they do not provide a unification of forces.

However, Alexander Polyakov had suggested that in certain cases the emergent strings associated with a gauge theory might behave like fundamental strings. Yet the gauge-theory strings would not exist in our world; instead, in one of the most remarkable feats of imagination in the history of the subject, Polyakov conjectured that they would move in a space that had one additional dimension.[5]

How did Polyakov succeed in conjuring up an extra dimension for his strings to move in? He found that when treated quantum-mechanically, the strings that arise from the gauge theory have an emergent property, which, it turns out, can be described by a number attached to each point on the string. A number can also be interpreted as a distance. In this case, Polyakov proposed that the number attached to each point of the string be interpreted as giving the position of that point in an additional dimension.

Taking this new emergent property into account, it was most natural to see the lines of electric flux of the field as living in a space with one more dimension. Thus, Polyakov was led to propose a duality between a gauge field in a world with three spatial dimensions and a string theory in a world with four spatial dimensions.

While Polyakov had made a general suggestion of this kind, it was Maldacena who refined the idea. In the world he studied, our three dimensions of space host the maximally super theory — the gauge theory with the maximal amount of supersymmetry. He studied the emergent strings that would arise as a dual description of that gauge theory. Extending Polyakov's argument, he found evidence that the string theory describing those emergent strings is actually a ten-dimensional supersymmetric string theory. Of the nine dimensions of space in which these strings live, four of them are like the ones in Polyakov's conjecture. There are, then, five dimensions left over, which are extra dimensions as described by Kaluza and Klein (see chapter 3). The extra five dimensions are arranged as a sphere. The four dimensions of Polyakov are curved, too, but in the opposite way from a sphere; such spaces are sometimes called saddle-shaped (see Fig. 12). These correspond to universes with dark energy, but where the dark energy is negative.

Maldacena's conjecture was much bolder than Polyakov's original proposal. It sparked an enormous response, and it has been the subject of thousands of papers written since. It has so far not been proved, but a great deal of evidence has accumulated that there is at least an approximate correspondence between string theory and gauge theory.

There was — and is — a great deal at stake here. If the Maldacena duality conjecture is right and the two theories are equivalent, then we have an exact quantum description of a quantum string theory. Any question we want to ask about the supersymmetric string theory can be translated into a question about the maximally super theory, which is a gauge theory. This is in principle much more than we had in other cases, where the string theory was defined at

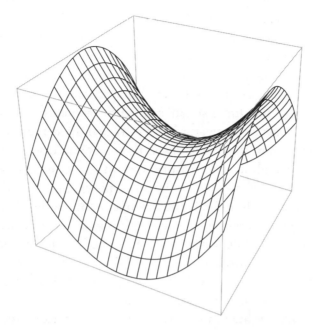

Fig. 12. A saddle-shaped surface, which is the geometry
of space in universes with negative energy density.

the background-dependent level only by a series of approximations.

There are, however, several caveats. Even if it is true, the duality conjecture can be useful only if one side of the duality can be defined precisely. So far, it has been possible to define the relevant version of string theory only in certain special cases. Thus, the hope was to go the other way and use the conjecture to define string theory in terms of the maximally super theory. However, while we knew much more about the maximally super theory, that theory also had not yet been precisely defined. There were hopes that we could do better, but they rested on difficult technical issues.

If Maldacena's conjecture is false, then the maximally super theory and string theory are not equivalent. However, even in this case, there is considerable evidence that at some levels of approximation there are useful relationships between the two. These approximations may not be strong enough to define one theory in terms of the other, but they do make it possible to calculate some properties of one relative to the other. A great deal of fruitful work along these lines has been done.

For example, at the lowest level of approximation, the ten-dimensional theory is just a version of general relativity extended to ten dimensions and enhanced by supersymmetry. This has no quantum mechanics and is well defined. It is easy to do some calculations in this theory, such as studying the propagation of different kinds of waves in the ten-dimensional spacetime geometry. Remarkably, even if Maldacena's conjecture proves true only at the lowest order of approximation, this has allowed us to calculate some properties of the corresponding gauge theory in our three-dimensional world.

This in turn led to insights into the physics of the other gauge theories. As a result, there is good evidence that, at least at the lowest level of approximation, string theories and gauge theories are related in the way Maldacena imagined. Whether the strong form of the Maldacena conjecture is true or false — indeed, even if string theory itself is false — we have gained powerful tools for understanding supersymmetric gauge theories.

After several years of intensive work, these matters remain confused. At issue is what exactly the relationship is between string theory and the maximally super theory. Most of the evidence is explained by a weak form of the Maldacena conjecture, which requires only that certain quantities in one theory be calculable using methods in the other and then only in a certain approximation. This, as I have noted, is already a result with important applications. But most string theorists believe in the strongest form of the conjecture, according to which the two theories are equivalent.

This situation is reminiscent of the strong-weak duality conjecture, in that it is possible to demonstrate the strongest results only on a very special subspace of states where there is a lot of extra symmetry. As in the strong-weak case, pessimists worried that the extra symmetry forced the theories to agree in a way they would not otherwise, whereas optimists were confident that the extra symmetry allowed us to achieve results that revealed a relationship that was true more generally.

Ultimately, it matters a lot which version of the Maldacena conjecture is true. One place it matters is in the description of black holes. Black holes can arise in universes with a negative dark energy, so one can try to use the Maldacena conjecture to study how the black-hole information paradox posed by Stephen Hawking is resolved. Depending on whether the correspondence between the two theories is exact or approximate, the resolution of the paradox could be different.

Suppose there is only a partial correspondence between the gravity theory in the interior of a black hole and the gauge theory. In that case, a black hole can trap information forever — or even pass the information on to a new universe born from the singularity at the center of a black hole, as some theorists, such as John Archibald Wheeler and Bryce DeWitt, long ago speculated. Thus the information is not lost after all, for it lives on in the new universe, but the information is lost forever to an observer at the black hole's boundary. This loss is possible if the gauge theory at the boundary contains only partial information about the interior. But suppose the correspondence between the two theories is exact. The gauge theory has neither horizons nor singularities and there is no place in which information can be lost. If it corresponds exactly to a spacetime with a black hole, no information can be lost there, either. In the first case, the observer loses information; in the second, he retains it. As of this writing, this issue has yet to be resolved.

As we have seen more than once, supersymmetry plays a fundamental role in string theory. String theories built without supersymmetry have instabilities; left alone, they will take off, emitting more and more tachyons in a process that has no end, until the theory breaks down. This is very unlike our world. Supersymmetry eliminates this behavior and stabilizes the theories. But in some respects, it does that too well. This is because supersymmetry implies that there is a symmetry in time, the upshot being that a supersymmetric theory cannot be built on a spacetime *that is evolving in time.* Thus, the aspect of the theory required to stabilize it also makes it difficult to study questions we would most like a quantum theory of gravity to answer, like what happened in the universe just after the Big Bang, or what happens deep inside the horizon of a black hole. Both are circumstances where the geometry is evolving rapidly in time.

This is typical of what we learned about string theory during the second superstring revolution. Our understanding expanded greatly, following a set of fascinating, unprecedented results. They gave us tantalizing hints of what might be true, if one could only peer behind an ever present veil and see the real thing. But try as we might, many of the calculations we wanted to do remained out of reach. To get any results, we had to choose special examples and conditions. In many instances, we were left not knowing whether the calculations we *could* do gave results that were a true guide to the general situation or not.

I personally found this situation very frustrating. We were either making fast progress toward the theory of everything, or we were off on a wild-goose chase, unwisely overinterpreting results, always taking the most optimistic reading from the calculations we were able to do. When I complained about this to some of the leaders of string theory in the mid 1990s, I was told not to worry, it was just that the theory was smarter than we were. We cannot, I was told, ask the theory questions directly and expect answers. Any direct attempt to solve the big problems was bound to fail. Instead, we should trust the theory and follow it, content to explore the parts it was willing to reveal using our imperfect methods of calculation.

There is only one catch. A genuine quantum version of M-theory would have to be background-independent, for the same reason that any quantum theory of gravity must be. But in addition to the reasons I spelled out earlier, M-theory must be background-independent because the five superstring theories, with all their different manifolds and geometries, are supposed to be part of M-theory. This includes all the different ways those geometries could be wrapped up, in all spatial dimensions from one to ten. They all provide backgrounds for strings and branes to move. But if they are part of one unified theory, that theory cannot be built on any one background, because it must encompass all backgrounds.

The key problem in M-theory, then, is to make a formulation of it that is consistent with quantum theory and background independence. This is an important issue, perhaps the most important open question in string theory. Unfortunately, not much progress has been made on it. There have been some fascinating hints, but we still do not know what M-theory is, or whether there is any theory deserving of the name.

There was some progress on an approach to a quantum-mechanical M-theory but, again, in a particular background. This was an attempt to make a quantum theory of the eleven-dimensional membrane theory, back in the 1980s. Three European physicists, Bernard de Wit, Jens Hoppe, and Hermann Nicolai, found that this could be done by a trick in which the membrane is represented by a two-dimensional table, or array, of numbers — called by mathematicians a *matrix*. Their formulation required that there be nine such tables, and from it they got a theory that approximates the behavior of the membrane.[8]

De Wit and his colleagues had found that you could make their matrix theory consistent with quantum theory. There was only one hitch, which was that to describe the membranes, the matrix had to extend infinitely, whereas the quantum theory could be shown to make sense only if the matrix was finite. So we were left with a conjecture that if the quantum theory could be consistently extended to infinite arrays of numbers, it would give a quantum theory of the membranes.

In 1996, four American string theorists revived this idea, but with a twist. Thomas Banks, Willy Fischler, Stephen Shenker, and Leonard Susskind proposed that on a background of eleven-dimensional flat spacetime, the same matrix theory gave not just the eleven-dimensional membrane theory but all of M-theory.[9] This matrix model doesn't give a full answer to what M-theory is, because it is in a particular background. It works in a few other backgrounds, but it cannot give sensible answers when more than four dimensions of space are wrapped up. If M-theory is right, our world has seven wrapped dimensions, so this isn't good enough. Moreover, we still don't know whether it leads to a completely consistent quantum theory if the matrix becomes infinite.

Unfortunately, M-theory remains a tantalizing conjecture. It's tempting to believe it. At the same time, in the absence of a real formulation, it is not really a theory — it is a conjecture about a theory we would love to believe in.

When I think of our relationship to string theory over the years, I am reminded of an art dealer who represented a friend of mine. When we met, he mentioned that he was also a good friend of a young writer whose book I had admired; we can call her "M." A few weeks later, he called me and said, "I was speaking to M. the other day, and, you know, she is very interested in science. Could I get you two together sometime?" Of course I was terribly flattered and excited and accepted the first of several dinner invitations. Halfway through a very good meal, the art dealer's cell phone rang. "It's M.," he announced. "She's nearby. She would love to drop by and meet you. Is that OK?" But she never came. Over dessert, the dealer and I had a great talk about the relationship between art and science. After a while, my curiosity about whether M. would actually show up lost out to my embarrassment over my eagerness to meet her, so I thanked him and went home.

A few weeks later he called, apologized profusely, and invited me to dinner again to meet her. Of course I went. For one thing, he ate only in the best restaurants; it seems that the managers of some art galleries have expense accounts that exceed the salaries of academic scientists. But the same scene was repeated that time and at several subsequent dinners. She would call, then an hour would go by, sometimes two, before his phone rang again: "Oh, I see, you're not feeling well" or "The taxi driver didn't know where the Odeon is? He took you to Brooklyn? What is this city coming to? Yes, I'm sure, very soon . . ." After two years of this, I became convinced that the picture of the young woman on her book jacket was a fake. One night I told him that I finally understood: He was M. He just smiled and said, "Well, yes . . . but she would have so enjoyed meeting you."

The story of string theory is like my forever postponed meeting with M. You work on it even though you know it's not the real thing, because it's as close as you know how to get. Meanwhile the company is charming and the food is good. From time to time, you hear that the real theory is about to be revealed, but somehow that never happens. After a while, you go looking for it yourself. This feels good, but it, too, never comes to anything. In the end, you have little more than you started with: a beautiful picture on the jacket of a book you can never open.

10

A Theory of <u>Anything</u>

IN THE TWO STRING revolutions, observation played almost no role. As the numbers of string theories grew, most string theorists continued to believe in the original vision of a unique theory that gave unique predictions for experiments, but there were no results that pointed in this direction, and a few theorists had worried all along that the unique theory would never emerge. Meanwhile, the optimists insisted that we must have faith and follow where the theory led. String theory appeared to do so much that was required of a unified theory that the rest of the story would surely reveal itself in time.

In the last several years, however, there has been a complete turnaround in how many string theorists think. The long-held hopes for a unique theory have receded, and many of them now believe that string theory should be understood as a vast landscape of possible theories, each of which governs a different region of a multiple universe.

What led to this reversal in expectations? Paradoxically, it was a confrontation with the data. But these were not the data we'd hoped for — they were data that most of us had never expected.

A good theory should surprise us; it means that whoever invented it was doing their job. But when an *observation* surprises us, theorists worry. No observation in the last thirty years has been more upsetting than the discovery of the dark energy in 1998. What we mean when we say that energy is dark is that it seems to differ from all forms of energy and matter previously known, in that it is not associated with any particles or waves. It is just there.

We do not know what the dark energy is; we know about it only because we can measure its effects on the expansion of the universe. It manifests itself as a source of gravitational attraction spread uniformly through space. Since it is distributed evenly, nothing falls toward it, for there is the same amount everywhere. The only effect it can have is on the average speed at which the galaxies move away from one another. What happened in 1998 was that observations of supernovas in distant galaxies indicated that the expansion of the universe was accelerating in a way that could best be explained by the existence of dark energy.[1]

One thing that the dark energy might be is something called the *cosmological constant.* This term refers to a form of energy with a remarkable feature: The properties of the energy, such as its density, appear exactly the same to all observers, no matter where they are in space and time and no matter how they are moving. This is highly unusual. Normally, energy is associated with matter, and there is a preferred observer, who moves with the matter. The cosmological constant is different. It is called a *constant* because you get the same universal value for it no matter where and when it is measured and how the observer is moving. Because it seems to have no origin or explanation in terms of particles or waves moving in space, it is called *cosmological* — that is, it is a feature of the whole universe and not any particular thing in it. (I should note that we are not yet sure that the dark energy is in fact in the form of a cosmological constant; all the evidence we have at present points that way, but we will know far better in the next few years whether the energy density is really unchanging in space and time.)

String theory did not predict the dark energy; even worse, the value detected was very hard for string theory to accommodate. Consequently, its discovery precipitated a crisis for the field. To understand why, we have to go back and tell the cosmological constant's strange, sordid story.

The story begins around 1916, with Einstein's refusal to believe the most dramatic prediction of his then new general theory of relativity. He had embraced the big lesson of general relativity, which was that the geometry of space and time evolves dynamically. So when people began applying his new theory to models of the universe, he should not have been surprised by what they found, which was that the universe, too, evolves dynamically in time. The model universes they studied expanded and contracted; they even seemed to have beginnings and ends.

But Einstein _was_ surprised by these results — and dismayed. From Aristotle up until that point, the universe had always been thought to be static. It might have been created by God, but if so, it hadn't changed since. Einstein was the most creative and successful theoretical physicist of the preceding two centuries, but even he could not imagine the universe as anything but eternal and immutable. We are tempted to say that if Einstein had been a real genius, he might have believed his theory more than his prejudice and predicted the expansion of the universe. But a more productive lesson is just how hard it is for even the most adventurous thinkers to give up beliefs that have been held for millennia.

We, who are so used to the idea now, can only speculate about how hard it was to accept the notion that the universe might have had a beginning. In any case, there was at the time no evidence that the universe changed or evolved in time, so Einstein took the predictions of an expanding universe as a sign that his theory was flawed and sought to reconcile it with his conception of an eternal universe.

He noticed that his equations for gravitation allowed a new possibility, which was that the energy density of empty space might have a value — in other words, it might not be zero. Furthermore, this universal energy density would be the same for all observers, no matter where or when they made observations, no matter how they moved. So he named it the cosmological constant. He found that the effect of the constant depended on its sign. When it was a positive number, it would cause the universe to expand — not only expand but do so at an accelerating rate. This is different from the effect of ordinary matter, which would cause the universe to contract because of the mutual gravitational attraction of all the matter it contains. So Einstein realized that he could use the expansive tendency of the new term to balance the contraction due to gravitational force, thus achieving a universe that was static and eternal.

Einstein later called the cosmological constant his biggest blunder. Actually, it was a blunder twice over. First, it didn't work very well; it didn't really keep the universe from contracting. You could balance the contraction caused by matter against the expansion caused by the cosmological constant, but only momentarily. The balance was inherently unstable. Tickle the universe and it would start to grow or shrink. But the real blunder was that the idea of a static universe was wrong to begin with. A decade later, an astronomer named Edwin Hubble began to find evidence that the universe was expanding. Since the 1920s, the cosmological constant has been an embarrassment, something to get rid of. But as time went on, this got harder and harder to do, at least theoretically. One could not just set it at zero and ignore it. Like the elephant in the corner, it was there even if you pretended it wasn't.

People soon began to understand that quantum theory had something to say about the cosmological constant. Unfortunately, it was the opposite of what we wanted to hear. Quantum theory — in particular, the uncertainty principle — appeared to require a huge cosmological constant. If something is exactly still, it has a definite position and momentum, and this contradicts the uncertainty principle, which says that you cannot know both these things about a particle. A consequence is that even when the temperature is zero, things keep moving. There is a small residual energy associated with any particle and any degree of freedom, even at zero temperature. This is called the *vacuum,* or *ground-state, energy.* When quantum mechanics is applied to a field, such as the electromagnetic field, there is a vacuum energy for every mode of vibration of the field. But a field has a huge number of modes of vibration; hence, quantum theory predicts a huge vacuum energy. In the context of Einstein's general theory of relativity, this implies a huge cosmological constant. We know this is wrong, because it implies that the universe would have expanded so fast that no structure at all could have formed. The fact that there are galaxies puts very strong limits on how big the cosmological constant can be. Those limits are some 120 orders of magnitude smaller than the predictions given by quantum theory; it might just qualify as the worst prediction ever made by a scientific theory.

Something is badly wrong here. A reasonable person could take the view that a radically new idea is needed and that no progress can be made in the unification of gravity and quantum theory until this discrepancy is explained. Several of the most sensible people feel this way. One of them is the German theoretical physicist Olaf Dreyer, who argues that the incompatibility between quantum theory and general relativity can be resolved only if we give up the idea that space is fundamental. He proposes that space itself emerges from a more fundamental description that is quite different. This point of view is also argued by several theorists who did great work in the field of condensed-matter physics, such as the Nobel laureate Robert Laughlin and the Russian physicist Grigori Volovik. But most of us who work on fundamental physics simply ignore this question and go on studying our different approaches, even if at the end of the day they do nothing to resolve it.

Until recently, there was a saving grace: At least the observed value of the cosmological constant was zero — that is, there was no evidence of an accelerating universal expansion rate. This was comforting, because we could hope that a new principle would be found that would eliminate the embarrassment from the equations altogether and make the cosmological constant exactly zero. It would have been far worse if the observed value had been some tiny nonzero number, because it's much harder to imagine a new principle cutting a number down to a much smaller but still non-zero number. Thus, for decades we thanked our various gods that at least we did not have that problem.

The cosmological constant posed a problem for all of physics, but the situation appeared a bit better for string theory. String theory could not explain why the cosmological constant was zero, but at least it explained why it was not a positive number. One of the few things we could conclude from the string theories then known was that the cosmological constant could only be zero or negative. I don't know of any particular string theorist who predicted that the cosmological constant could not be a positive number, but it was widely understood to be a consequence of string theory. The reasons are too technical to do justice to them here.

In fact, string theories with negative cosmological constants had been studied. The famous Maldacena conjecture, for example, involved a spacetime with a negative cosmological constant. There were a number of difficulties, and to this day no one has explicitly written down the details of a string theory in a world with a negative cosmological constant. But this lack of explicitness is believed to be a technical issue — there is no known reason why it should not be possible in principle.

You can imagine the surprise, then, in 1998, when the observations of supernovas began to show that the expansion of the universe was accelerating, meaning that the cosmological constant had to be a positive number. This was a genuine crisis, because there appeared to be a clear disagreement between observation and a prediction of string theory. Indeed, there were theorems indicating that universes with a positive cosmological constant — at least, as long as quantum effects were neglected — could not be solutions of string theory.

Edward Witten is not someone given to pessimism, yet he flatly declared in 2001 that "I don't know any clear-cut way to get de Sitter space [a universe with a positive cosmological constant] from string theory or M-theory."[2]

Philosophers and historians of science, among them Imre Lakatos, Paul Feyerabend, and Thomas Kuhn, have argued that one experimental anomaly is rarely enough to kill a theory. If a theory is believed deeply enough, by a large enough group of experts, they will go to ever more extreme measures to save it. This is not always bad for science, and occasionally it can be very good. Sometimes the theory's defenders succeed, and when they do, great and unexpected discoveries can be made. But sometimes they fail, and then lots of time and energy is wasted as theorists dig themselves deeper and deeper into a hole. The story of string theory in the last few years is one that Lakatos or Feyerabend would have understood well, for it is the story of a large group of experts doing what they can to save a cherished theory in the face of data that seem to contradict it.

What saved string theory — if indeed it has been saved — was the solution to an entirely different problem: how to make the higher dimensions stable. Recall that in higher-dimensional theories the curling up of the extra dimensions produces many solutions. The ones that could possibly reproduce the world we observe are very special, in that certain aspects of the geometry of the higher-dimensional spaces have to be kept frozen. Otherwise, once the geometry starts to evolve, it may just keep going, resulting in either a singularity or a fast expansion that makes the curled-up extra dimensions as big as the dimensions we observe.

String theorists called this the *problem of moduli stabilization,* "moduli" being a general name for the constants that denote the properties of the extra dimensions. This was a problem that string theory had to solve, but for a long time it was not clear how. As in other cases, the pessimists fretted, while the optimists were confident that sooner or later we'd discover the solution.

In this case, the optimists were right. Progress began in the 1990s, when several theorists in California understood that the key was to use the branes to stabilize the higher dimensions. To understand how, we have to appreciate one feature of the problem, which is that the geometry of the higher dimensions can vary continuously while remaining a good background for a string theory. In other words, you can vary the volume or the shape of the higher dimensions and, by doing so, have them flow through a space of different string theories. This means that there is nothing to stop the geometry of the extra dimensions from evolving in time. To avoid this evolution, we had to find a class of string theories that were impossible to move seamlessly among. One way to do this was to find string theories for which every change is a discrete step — that is, instead of flowing smoothly among theories, you have to make big, abrupt changes.

Joseph Polchinski told us that there were indeed discrete objects in string theory: branes. Recall that there are string backgrounds in which branes are wrapped around surfaces in the extra dimensions. Branes come in discrete units. You can have 1, 2, 17, or 2,040,197 branes but not 1.003 branes. Since branes carry electric and magnetic charges, this gives rise to discrete units of electric and magnetic flux.

So in the late 1990s, Polchinski, working with an imaginative postdoc named Raphael Bousso, began to study string theories in which large numbers of units of electric flux are wrapped around the extra dimensions. They were able to get theories in which some parameters could no longer vary continuously.

But could you freeze all the constants this way? This required a much more complicated construction, but the answer had an added benefit. *It made a string theory with a positive cosmological constant.*

The crucial breakthrough was made in early 2003, by a group of scientists from Stanford, including Renata Kallosh, a pioneer of supergravity and string theory; Andrei Linde, who is one of the discoverers of inflation; and two of the best young string theorists, Shamit Kachru and Sandip Trivedi.[3] Their work is complicated even by the standards of string theory; it has been characterized by their Stanford colleague Leonard Susskind as a "Rube Goldberg contraption." But it had a huge impact, because it solved both the problem of stabilizing the extra dimensions and the problem of making string theory consistent with the observations of dark energy.

Here's a simplified version of what the Stanford group did. They started with a much-studied kind of string theory — a flat four-dimensional spacetime with a small six-dimensional geometry over each point. They chose the geometry of the six wrapped-up dimensions to be one of the Calabi-Yau spaces (see chapter 8). As noted, there are at least a hundred thousand of these, and all you have to do is pick a typical one whose geometry depends on many constants.

Then they wrapped large numbers of electric and magnetic fluxes around the six-dimensional spaces over each point. Because you can wrap only discrete units of flux, this tends to freeze out the instabilities. To further stabilize the geometry, you have to call on certain quantum effects not known to arise directly from string theory, but they are understood to some extent in supersymmetric gauge theories, so it is possible that they play a role here. Combining these quantum effects with the effects from the fluxes, you get a geometry in which all the moduli are stable.

This can also be done so that there appears to be a negative cosmological constant in the four-dimensional spacetime. It turns out that the smaller we want the cosmological constant to be, the more fluxes we must wrap, so we wrap huge numbers of fluxes to get a cosmological constant that is tiny but still negative. (As noted, we don't know explicitly how to write the details of a string theory on such a background, but there's no reason to believe it doesn't exist.)

But the point is to get a *positive* cosmological constant, to match the new observations of the universe's expansion rate. So the next step is to wrap other branes around the geometry, in a different way, which has the effect of raising the cosmological constant. Just as there are antiparticles, there are antibranes, and the Stanford group used them here. By wrapping antibranes, energy can be added so as to make the cosmological constant small and positive. At the same time, the tendency of string theories to flow into one another is suppressed, because any change requires a discrete step. Thus, two problems are solved at once: The instabilities are eliminated and the cosmological constant is small and positive.

The Stanford group may have saved string theory, at least for the time being, from the crisis generated by the cosmological constant. But the way they did it had such weird and unintended consequences that it has split the string community into factions. Before this, the community had been remarkably in accord. Going to a string theory conference in the 1990s was like going to China in the early 1980s, in that almost everyone you talked with seemed to fervently hold the same point of view. For better or worse, the Stanford group destroyed the party unity.

Recall that the particular string theory we are discussing comes from wrapping fluxes around the compact geometries. To get a small cosmological constant, you have to wrap many fluxes. But there's more than one way to wrap a flux; in fact, there are a lot of ways. How many?

Before answering this question, I have to emphasize that we don't know if *any* of the theories made by wrapping fluxes around the hidden dimensions give good consistent quantum string theories. The question is too hard to answer using the methods we have. So what we do is apply tests, which give us *necessary but insufficient* conditions for good string theories to exist. The tests require that the string theories, if they exist, have strings that interact weakly. This means that if we could do calculations in the string theories, the results would be very close to predictions of the approximate calculations we are able to do.

A question we *can* answer is how many string theories pass these tests, which involve wrapping fluxes around the six hidden dimensions. The answer depends on what value of the cosmological constant we want to come out. If we want to get a negative or zero cosmological constant, there are *an infinite number of distinct theories*. If we want the theory to give a positive value for the cosmological constant, so as to agree with observation, there are a finite number; at present there is evidence for 10^{500} or so such theories.

This is of course an enormous amount of string theories. Moreover, each one is distinct. Each will give different predictions for the physics of the elementary particles and different predictions for the values of the parameters of the standard model.

The idea that string theory gave us not one theory but a landscape consisting of many possible theories had been proposed in the late 1980s and early 1990s, but it had been rejected by most theorists. As noted, Andrew Strominger had found in 1986 that there was a huge number of apparently consistent string theories, and a few string theorists had continued to worry about the resulting loss of predictivity, while most of them had remained confident that a condition would emerge that would settle on a unique and correct theory. But the work of Bousso and Polchinski and the Stanford group finally tipped the balance. It gave us an enormous number of new string theories, as Strominger had, but what was new was that these numbers were needed to solve two big problems: that is, to make string theory consistent with the observations of a positive vacuum energy and to stabilize the theories. Probably for these reasons, the vast landscape of theories finally came to be seen not as a freak result to be ignored but as a means of saving string theory from being falsified.

Another reason the landscape idea took hold was, quite simply, that theorists were discouraged. They had spent a long time searching for a principle that would select a unique string theory, but no such principle had been discovered. Following the second revolution, string theory was now much better understood. The dualities, in particular, made it more difficult to argue that most string theories would be unstable. Thus, string theorists began to accept the vast landscape of possibilities. The question driving the field was no longer how to find a unique theory but how to do physics with such a huge collection of theories.

One response is to say that it's impossible. Even if we limit ourselves to theories that agree with observation, there appear to be so many of those that some of them will almost certainly give you the outcome you want. Why not just take this situation as a _reductio ad absurdum_? That sounds better in Latin, but it's more honest in English, so let's say it: If an attempt to construct a unique theory of nature leads instead to 10^{500} theories, that approach has been reduced to absurdity.

This is painful for many who have invested years and even decades of their working lives in string theory. If it is painful for me, having devoted a certain amount of time to the effort, I can only imagine how some of my friends who have staked their whole careers on string theory must feel. Still, even if it hurts like hell, acknowledging the _reductio ad absurdum_ seems a rational and honest response to the situation. It is a response that a few people I know have chosen. But it is not one that most string theorists choose.

There is another rational response: Deny the claim that a vast number of string theories exist. The arguments for the new theories with positive cosmological constant are based on drastic approximations; perhaps they lead theorists to believe in theories that do not exist mathematically, let alone physically.

In fact, the evidence for a vast number of string theories with a positive cosmological constant is based on very indirect arguments. We do not know how to actually describe strings moving in these backgrounds. Moreover, we can define some necessary conditions for a string theory to exist, but we don't know whether these conditions are also _sufficient_ for the theory's existence. There is, then, no proof that a theory of strings really exists in any of these backgrounds. So a rational person might say that perhaps they don't. Indeed, there are recent results — from Gary Horowitz, who is one of the discoverers of the Calabi-Yau spaces, and two younger colleagues, Thomas Hertog and Kengo Maeda — that raise questions about whether _any_ of these theories describe stable worlds.[4] One can either take such evidence seriously or ignore it, which is what many string theorists are doing. The possible instability found by Horowitz and his collaborators afflicts not just the landscape of new theories found by the Stanford group but all solutions that involve the six-dimensional Calabi-Yau spaces. If these solutions are indeed all unstable, it means that most of the work aimed at connecting string theory with the real world will have to be thrown out. There is also currently a debate over the validity of some of the Stanford group's assumptions.

At the beginning of the first superstring revolution, it was miraculous that any string theory existed at all. That there were eventually five was even more surprising. The sheer improbability cemented our belief in the project. If at first it was unlikely to work and then it did work — well, this was nothing less than wonderful. Today string theorists are ready to accept the existence of a landscape containing a vast number of theories, based on much less evidence than we needed twenty years ago to convince ourselves that a single theory existed.

So one place to draw the line is simply to say, "I need to be persuaded that these theories exist, using the same standards that were required decades ago to evaluate the original five." If you insist on those standards, then you will not believe in the vast number of new theories, because the evidence for any theory in the current landscape is pretty minimal according to the old standards. This is the point of view I find myself leaning toward, most of the time. It just seems to me the most rational reading of the evidence.

11

The Anthropic Solution

MANY PHYSICISTS I know have lowered their expectations that string theory is the fundamental theory of nature — but not everyone. In the last few years, it has become fashionable to argue that the problem lies not with string theory but with our expectations of what any physical theory should look like. That argument was introduced a couple of years ago by Leonard Susskind in a paper called "The Anthropic Landscape of String Theory":

> Based on the recent work of a number of authors, it seems plausible that the landscape is unimaginably large and diverse. Whether we like it or not, this is the kind of behavior that gives credence to the Anthropic Principle. . . . The [theories in the Stanford group's landscape] are not at all simple. They are jury-rigged, Rube Goldberg contraptions that could hardly have fundamental significance. But in an anthropic theory simplicity and elegance are not considerations. The only criteria for choosing a vacuum is utility, i.e. does it have the necessary elements such as galaxy formation and complex chemistry that are needed for life. That together with a cosmology that guarantees a high probability that at least one large patch of space will form with that vacuum structure is all we need.[1]

The anthropic principle that Susskind refers to is an old idea proposed and explored by cosmologists since the 1970s, dealing with the fact that life can arise only in an extremely narrow range of all possible physical parameters and yet, oddly enough, here we are, as though the universe had been designed to accommodate us (hence the term "anthropic"). The specific version that Susskind invokes is a cosmological scenario that has been advocated by Andrei Linde for some time, called *eternal inflation.* According to this scenario, the rapidly inflating phase of the early universe gave rise not to one but to an infinite population of universes. You can think of the primordial state of the universe as a phase that is exponentially expanding and never stops. Bubbles appear in it, and in these places the expansion slows dramatically. Our world is one of those bubbles, but there are an infinite number of others. To this scenario, Susskind adds the idea that when a bubble forms, one of the vast number of string theories is chosen by some natural process to govern that universe. The result is a vast population of universes, each of which is governed by a string theory randomly chosen from the landscape of theories. Somewhere in the so-called multiverse is every possible theory in the landscape.

I find it unfortunate that Susskind and others have embraced the anthropic principle, because it has been understood for some time that it is a very poor basis for doing science. Since every possible theory governs some part of the multiverse, we can make very few predictions. It is not hard to see why.

To make a prediction in a theory that posits a vast population of universes satisfying randomly chosen laws, we would first have to write down all the things we know about our own universe. These things would apply to some number of other universes as well, and we can refer to the subset of universes where these facts are true as *possibly true universes.*

All we know is that our universe is one of the possible universes. Given that the population of universes was produced by randomly distributing the fundamental laws of nature among them, we can know little else. We can make a new prediction only if every, or almost every, possibly true universe has a property that is not on the list of the properties we've already observed in our own universe.

For example, suppose that in almost every possibly true universe the most resonant oscillation is a low C. Then it is highly probable that a universe picked randomly from the possibly true universes will be resonating at low C. Since we can know nothing about our own universe except that it is a possibly true universe, we can predict with a high probability that our universe is singing a low C, too.

The problem is that since the distribution of theories over all the universes is assumed to be random, there are very few properties like this. Most likely, once we have specified the properties we observe in our own universe, the remaining properties that any universe might have will be distributed randomly among the other possibly true universes. Thus we can make no predictions.

What I have been describing is what cosmologists call the *weak anthropic principle.* As the name indicates, the one thing we know about our universe is that it supports intelligent life; therefore, every possibly true universe must be a place where intelligent life could live. Susskind and others argue that this principle is nothing new. For example, how do we explain the fact that we find ourselves on a planet situated so that the temperature is in the range where water is liquid? If we believed that there was only one planet in the universe, we would find this fact puzzling. We would be tempted to believe in the necessity of an intelligent designer. But once we know that there are a vast number of stars and many planets, we understand that just by chance there will be many planets friendly to life. Therefore we are not surprised to find ourselves on one of them.

There is, however, a big difference between the planet analogy and the cosmological situation, which is that we do not know of any universes except our own. The existence of a population of other universes is a hypothesis that cannot be confirmed by direct observation; hence, it cannot be used in an explanatory fashion. It is true that *if* there were a population of universes with random laws, we would not be surprised to find ourselves in one where we can live. But the fact that we are in a biofriendly universe cannot be used as a confirmation of a theory that there is a vast population of universes.

There is a counterargument, which we can illustrate with the planet example. Let us suppose that it was impossible to observe any other planets. If we deduced from this that there was in fact only one planet, we would be forced to believe in something very improbable, which is that the single planet that exists is biofriendly. On the other hand, if we assume that there are many planets with random properties, even if we can never observe them, then the probability that a few are biofriendly is greatly increased — in fact, it approaches 1. Therefore, it is argued, it is much more probable that there are many planets than that there is only one.

But this apparently strong argument is fallacious.* To see why, let us compare it to another argument that might be made from the same evidence. Someone who believed in intelligent design could argue that if there is only one planet and it is biofriendly, there is a high probability of an intelligent designer at work. Given a choice between two theories — (1) the unique planet is biofriendly just by extreme luck, and (2) there was an intelligent designer who made the unique planet and made it biofriendly — the same logic leads us to conclude that it is more rational to choose the second alternative.

The scenario of many unobserved universes plays the same logical role as the scenario of an intelligent designer. Each provides an untestable hypothesis that, if true, makes something improbable seem quite probable.

Part of the reason these arguments are fallacious is that they rely on an unstated assumption — that we have in hand the complete list of alternatives. For (to return to the planet analogy) we cannot preclude the possibility that a genuine explanation for our planet's biofriendliness will emerge sometime in the future. The fallacy in the two arguments is that they both compare a single possible — but untestable — explanation with the statement that there is no possible explanation. Of course, given only those two choices, an explanation appears more rational than any unexplained improbability.

For centuries we have had good reason to believe there are lots of planets, because there are lots of stars — and recently we have confirmed directly the existence of extrasolar planets. So we believe the many-planet explanation for the biofriendliness of our planet. But when it comes to the biofriendliness of our universe, we have at least three possibilities:

1. Ours is one of a vast collection of universes with random laws.
2. There was an intelligent designer.

* The fallacious principle used here goes like this. Let us observe O and consider two explanations. Given explanation A, the probability of O is very low, but given explanation B, the probability is high. It is tempting to deduce from this that the probability is much higher for B than for A, but there is no principle of logic or probability that allows this deduction.

3. There is a so-far-unknown mechanism that will both explain the biofriendliness of our universe and make testable predictions by which it can be confirmed or falsified.

Given that the first two possibilities are untestable in principle, it is most rational to hold out for the third possibility. Indeed, that is the only possibility we should consider as scientists, because accepting either of the first two would mean the end of our field.

Some physicists claim that the weak anthropic principle must be taken seriously because it has led in the past to genuine predictions. I'm talking here about some of the people I most admire — not only Susskind but also Steven Weinberg, the physicist who, you'll recall from chapter 4, together with Abdus Salam unified the electromagnetic and weak nuclear forces. It then pains me to conclude that in every case I've looked into, the claims have proved erroneous.

Consider, for example, the following argument about the properties of the carbon nuclei, based on investigations made in the 1950s by the great British astrophysicist Fred Hoyle. This argument is often taken as indicating that real physical predictions can be based on the anthropic principle. The argument begins with the observation that for life to exist, there must be carbon. Indeed, carbon is plentiful. We know that it could not have been made in the Big Bang, hence we know that it must have been made in stars. Hoyle noticed that carbon could have been formed in stars only if there was a certain resonant state of carbon nuclei. He communicated this prediction to a group of experimentalists, who found it.

The success of Hoyle's prediction is sometimes used as support for the effectiveness of anthropic principle. But the argument from life in the preceding paragraph has no logical relation to the rest of the paragraph. What Hoyle did was to reason from the observation that the universe is full of carbon to a conclusion based on the necessity of there being some process whereby all that carbon got made. The fact that we and other living things are made of carbon is unnecessary to the argument.

Another example often cited in support of the anthropic principle is a prediction about the cosmological constant made in a celebrated 1987 paper by Steven Weinberg. In it, he pointed out that the cosmological constant must be less than a certain value, otherwise the universe would have expanded too rapidly for galaxies to form.[2] Since we observe that the universe is full of galaxies, the cosmological constant must be less than that value. And it is, as it must be. This is perfectly good science. But Weinberg took this valid scientific argument further. Suppose there is a multiverse, he said, and suppose the values of the cosmological constant are randomly distributed among its member universes. Then, among the possibly true universes, a typical value of the cosmological constant would be of the order of magnitude of the largest that is consistent with galaxy formation. Hence, if the multiverse scenario is true, we should expect the cosmological constant to be as large as it can be, while still allowing galaxies to form.

When Weinberg published this prediction, the cosmological constant was generally believed to be zero. Thus it was impressive that his prediction came true to within roughly a factor of 10. However, when the new results forced Weinberg's statements to be examined more carefully, some problems emerged. Weinberg had considered a population of universes in which only the cosmological constant is randomly distributed, while all the other parameters are held fixed. Instead, he should have averaged over all the members of the multiverse consistent with galaxy formation, allowing *all* the parameters to vary. Done this way, the prediction of the cosmological constant's value turns out to be much further off.

This illustrates a persistent problem with reasoning of this type. If your scenario invokes randomly distributed parameters, of which you can observe only one set, you can get a wide range of possible predictions, depending on the precise assumptions you make about that unknown, unobservable population of other sets. For example, each of us is a member of many communities. In many of these we will be typical members, but in many others we will be untypical. Suppose, in my author's bio on the book jacket, all I write is that I am a typical person. How much about me will you be able to deduce?

There are many other cases in which some version of the weak anthropic principle may be tested. Within the standard model of elementary-particle physics, there are constants that simply don't have the values we would expect them to have if they were chosen by random distribution among a population of possibly true universes. We would expect that the quark and lepton masses, except for the first generation, would be randomly distributed, but relations between them are seen. We would expect that some symmetries of the elementary particles would be violated by the strong nuclear force much more than they are. We would expect the proton to decay at a much faster rate than present experimental limits allow. In fact, I know of no successful predictions that have been made by reasoning from a multiverse with a random distribution of laws.

But what about the third possibility, which is an explanation for the biofriendliness of our universe based on testable hypotheses? In 1992 I put a proposal of just this kind on the table. To get testable predictions from a multiverse theory, the population of universes must be far from random. It must be intricately structured so that there are properties that all or most universes have that have nothing to do with our existence. We can then predict that our universe has these properties.

One way to get such a theory is to mimic the way natural selection works in biology. I invented such a scenario in the late 1980s, when it became clear that string theory would come in a very large number of versions. From books by evolutionary biologists Richard Dawkins and Lynn Margulis, I learned that biologists had models of evolution that were based on a space of possible phenotypes they called *fitness landscapes*. I adopted the idea and the term and invented a scenario in which universes are born from the interiors of black holes. In *The Life of the Cosmos* (1997), I reflected at length on the implications of this idea, so I will not go into it in detail here, except to say that that theory, which I called *cosmological natural selection*, made genuine predictions. In 1992 I published two of them and they have since held up, although they could have been proved false by many observations made since then. These are (1) that there should be no neutron stars more massive than 1.6 times the mass of the sun, and (2) that the spectrum of fluctuations generated by inflation — and, plausibly, observed in the cosmic microwave background — should be consistent with the simplest possible version of inflation, with one parameter and one inflaton field.[3]

Susskind, Linde, and others have attacked the idea of cosmological natural selection, because they claim that the multitudes of universes created in eternal inflation will overwhelm any numbers made through black holes. To address this objection, it is important to know how reliable the prediction of eternal inflation is. The case is sometimes made that it is hard to have inflation at all without eternal inflation. The fact that some of the predictions of inflationary cosmology have been confirmed is taken as evidence for it. However, moving from inflation to eternal inflation assumes that there is no barrier to extending conclusions that hold on our present cosmological scale to vastly bigger scales. There are two problems with this: The first is that the extrapolation to bigger scales at the present time implies, in some models of inflation, an extrapolation to much smaller scales in the early universe. (I will not explain this here, but it is true of several inflation models.) This means that to get an inflated universe vastly bigger than our present universe, we must extend the description of the early universe to times vastly smaller than the Planck time, before which quantum gravity effects dominate the evolution of the universe. This is problematic, because the usual description of inflation assumes that spacetime is classical and there are no effects of quantum gravity; moreover, several theories of quantum gravity predict that there is no interval of time shorter than the Planck time. Second, there are indications that the predictions of inflation are not satisfied on the largest scales we can currently observe (see chapter 13). Hence, the extrapolation from inflation to eternal inflation runs into trouble both theoretically and observationally, so this does not seem a strong objection to cosmological natural selection.

In spite of the fact that the anthropic principle has not led to any real predictions and is not likely to, Susskind, Weinberg, and other leading theorists have embraced it as signaling a revolution not just in physics but in our conception of what a physical theory is. Weinberg asserts in a recent essay,

> Most advances in the history of science have been marked by discoveries about nature, but at certain turning points we have made discoveries about science itself. . . . Now we may be at a new turning point, a radical change in what we accept as a legitimate foundation for a physical theory. . . . The larger the number of possible values of physical parameters provided by the string landscape, the more string theory legitimates anthropic reasoning as a new basis for physical theories: Any scientists who study nature must live in a part of

the landscape where physical parameters take values suitable for the appearance of life and its evolution into scientists.[4]

Steven Weinberg is justly honored for his contributions to the standard model, and his writings usually advance a compelling and sober rationality. But, simply put, once you reason like this, you lose the ability to subject your theory to the kind of test that the history of science shows over and over again is required to winnow correct theories from beautiful but wrong ones. To do this, a theory must make specific and precise predictions that can either be confirmed or refuted. If there is a high risk of disconfirmation, then confirmation counts for a lot. If there is no risk of either, then there is no way to continue to do science.

The debate about how science is to confront the newly vast string landscape seems to me to come down to three possibilities:

1. String theory is right, and the random multiverse is right, so to accommodate them we must change the rules that govern how science works, because according to the usual scientific ethic, we would not allow ourselves to believe in a theory that made no unique predictions by which it could be confirmed or falsified.
2. Some way will eventually be found to deduce genuine unique and testable predictions from string theory. This might be done either by showing that there really is a unique theory or by a different, nonrandom multiverse theory that leads to genuine testable predictions.
3. String theory is not the right theory of nature. Nature is best described by another theory, yet to be discovered or yet to be accepted, which does lead to genuine predictions, which experiments will eventually confirm.

What is remarkable to me is the number of distinguished scientists who seem unable to accept the possibility either that string theory or the hypothesis of a random multiverse is wrong. Here is a collection of pertinent comments:

"Anthropic reasoning runs so much against the historic goals of theoretical physics that I resisted it long after realizing its likely necessity. But now I have come out." — JOSEPH POLCHINSKI

"Those who dislike the anthropic principle are simply in denial."

— ANDREI LINDE

"The possible existence of a huge landscape is a fascinating development in theoretical physics that forces a radical rethinking of many of our assumptions. My gut feeling is that it may well be right."

— NIMA ARKANI-HAMED (Harvard University)

"I think it's quite plausible that the landscape is real."

— MAX TEGMARK (MIT)

Even Edward Witten seems stumped: "I just don't have anything incisive to say. I hope we will learn more."[5]

There is not a person quoted here whom I do not deeply admire. Nevertheless, it seems to me that any fair-minded person not irrationally committed to a belief in string theory would see this situation clearly. A theory has failed to make any predictions by which it can be tested, and some of its proponents, rather than admitting that, are seeking leave to change the rules so that their theory will not need to pass the usual tests we impose on scientific ideas.

It seems rational to deny this request and insist that we should not change the rules of science just to save a theory that has failed to fulfill the expectations we originally had for it. If string theory makes no unique predictions for experiments, and if it explains nothing about the standard model of particle physics which was previously mysterious — apart from the obvious statement that we must live in a universe where we can live — it does not seem to have turned out to be a very good theory. The history of science has seen a lot of initially promising theories fail. Why is this not another such case?

We have regrettably reached the conclusion that string theory has made no new, precise, and falsifiable predictions. But still, string theory makes some startling assertions about the world. Could an experiment or an observation one day reveal evidence for any of these surprising features? Even if there are no definite up or down predictions — predictions of the kind that could kill or confirm the theory — might we see evidence of a feature that is central to the stringy view of nature?

The most obvious novelty of string theory is the strings themselves. If we could probe the string scale, there would be no problem seeing abundant evidence for string theory, if it is true. We would see indications that the fundamental objects are one-dimensional rather than pointlike. But we are not able to do accelerator experiments at anywhere near the energies required. Is there another way we could make the strings reveal themselves? Might the strings somehow be induced to become bigger, so that we could see them?

One such scenario was proposed recently by Edmund Copeland, Robert Myers, and Joseph Polchinski. Under certain very special assumptions about cosmology, it might be true that some very long strings were created in the early universe and continue to exist.[6] The expansion of the universe has stretched them to the point that they now are millions of light-years long.

This phenomenon is not limited to string theory. For some time, a popular theory about the formation of galaxies suggested that they were seeded by the presence of huge strings of electromagnetic flux left over from the Big Bang. These *cosmic strings,* as they were called, had nothing to do with string theory; they were a consequence of the structure of the gauge theories. They are analogous to the quantized lines of magnetic flux in superconductors, and they can form in the early universe as a consequence of the universe going through phase transitions as it cools down. We now have definitive evidence from cosmological observations that such strings were not the main ingredient in the formation of structure in the universe, but there could still be some cosmic strings left over from the Big Bang. Astronomers search for them by looking for their effect on light from distant galaxies. If a cosmic string were to come between our line of sight and a distant galaxy, the gravitational field of the string would act as a lens, duplicating the image of the galaxy in characteristic ways. Other objects, like dark matter or another galaxy, can have a similar effect, but astronomers know how to distinguish between the images they generate and those produced by a cosmic string. Recently there was a report that such a lens might have been detected. It was labeled, optimistically, CSL-1, but when it was viewed by the Hubble Space Telescope, it turned out to be two galaxies close to each other.[7]

What Copeland and his colleagues found is that under certain special conditions, a fundamental string, stretched to enormous lengths by the expansion of the universe, would resemble a cosmic string. So it might be observable through its action as a lens. Such a fundamental cosmic string may also be a prodigious radiator of gravitational waves, which might make it observable by LIGO, the Laser Interferometer Gravitational-wave Observatory.

Predictions of this kind give us some hope that string theory might someday be verified by observations. Yet the discovery of a cosmic string, by itself, cannot verify string theory, because several other theories also predict the existence of such strings. Nor can failure to find one lead to a falsification of string theory, because the conditions for such cosmic strings to exist are specially chosen, and there is no reason to think they might exist in our universe.

Besides the existence of strings, there are three other generic features of the stringy world. All sensible string theories agree that there are extra dimensions, that all the forces are unified into one force, and that there is supersymmetry. So even if we have no detailed predictions, we can see whether experiment can test these hypotheses. Since they are independent of string theory, finding evidence for any one of them does not prove that string theory is true. But here the opposite is not the case: If we learn that there is no supersymmetry or no higher dimensions or no unification of all the forces, then string theory is false.

Let's start with extra dimensions. We may not be able to see them, but we can certainly look for their effects. One way to do this is to search for the extra forces that are predicted by all higher-dimensional theories. These forces are transmitted by the fields that comprise the geometry of the extra dimensions. Such fields must be there, because you cannot limit the extra dimensions to producing only the fields and forces we so far observe.

The forces that come from such fields are expected to be roughly as strong as gravity, but they may differ from gravity in one or more ways: They may have a finite range, and they may not interact equally with all forms of energy. Some current experiments are extraordinarily sensitive to such hypothetical forces. About ten years ago, one experiment saw preliminary evidence for such a force, which was called the *fifth force.* Further experiments did not support the claim, and as of yet there is no evidence for such forces.

String theorists have usually assumed that the extra dimensions are tiny, but several adventurous physicists realized in the 1990s that this did not have to be the case — that the extra dimensions could be large or even infinite. This is possible in a brane-worlds scenario. In such a picture, our three-dimensional space is actually a brane — that is, something like a physical membrane but with three dimensions — suspended in a world with four or more dimensions of space. The particles and forces of the standard model — electrons, quarks, photons, and the forces they interact with — are restricted to the three-dimensional brane that makes up our world. So using only those forces, you cannot see evidence of the extra dimensions. The sole exception is the gravitational force. Gravity, being universal, extends through all the dimensions of space.

This kind of scenario was first constructed in detail by three physicists working at SLAC, the Stanford Linear Accelerator Center: Nima Arkani-Hamed, Gia Dvali, and Savas Dimopoulos. Surprisingly, they found that the extra dimensions could be quite large without conflicting with known experiments. If there were two extra dimensions, they could be as wide as a millimeter across.[8]

The main effect of adding such large extra dimensions is that the gravitational force in the four- or five-dimensional world turns out to be much stronger than it appears to be on the three-dimensional brane, so quantum gravitational effects happen at a much larger length scale than otherwise expected. In quantum theory, a larger length scale means a smaller energy. By making the extra dimensions as large as a millimeter, one can bring down the energy scale at which quantum-gravity effects should be seen — from the Planck energy, which is 10^{19} GeV, to only 1,000 GeV. This would resolve one of the most stubborn questions about the parameters of the standard model: that is, Why is the Planck energy so many orders of magnitude bigger than the mass of the proton? But what is really exciting is that it would bring quantum-gravity phenomena within range of being revealed by the Large Hadron Collider (LHC), coming online in 2007. Among these effects could be the production of quantum black holes in collisions of elementary particles. This would be a dramatic discovery.

Another kind of brane-world scenario was developed by Lisa Randall, of Harvard, and Raman Sundrum, of Johns Hopkins University.

They found that the extra dimensions could be infinite in size as long as there was a negative cosmological constant in the higher-dimensional world.[9] Remarkably, this too agrees with all observations to date, and it even makes predictions for new ones.

These are adventurous ideas and fun to think about, and I'm deeply admiring of their inventors. That said, I'm troubled by the brane-world scenarios. They are vulnerable to the same problems that doomed the original attempts at unification through higher dimensions. The brane-world scenarios work only if you make special assumptions about the geometry of the extra dimensions and the way the three-dimensional surface that is our world sits inside them. In addition to all the problems suffered by the old Kaluza-Klein theories, there are new problems. If there can be one brane floating in the higher-dimensional world, couldn't there be many? And if there are others, how often do they collide? Indeed, there are proposals that the Big Bang arose from the collision of brane worlds. But if this can happen once, why hasn't it happened since? Some 14 billion years have gone by. The answer might be that branes are scarce, in which case we are back to depending on very finely tuned conditions. Or it might be that branes are precisely parallel to one another and don't move much, in which case we again have finely tuned conditions.

Beyond these problems, I am skeptical because these scenarios depend on special choices of background geometries, and this contradicts Einstein's principal discovery, as set out in his general theory of relativity, that the geometry of spacetime is dynamical and that physics must be expressed in a background-independent manner. Nevertheless, this is science as it should be: bold ideas that are testable by doable experiments. Let's be clear, though. If any of the predictions of brane worlds comes true, it will not amount to a confirmation of string theory. Brane-world theories stand on their own; they do not need string theory. Nor is there a completely worked-out realization of a brane-world model within string theory. Conversely, if none of the predictions of brane worlds are seen, this does not falsify string theory. Brane worlds are just one of the ways that the extra dimensions of string theory could manifest themselves.

The second generic prediction of string theory is that the world is supersymmetric. Here, too, there is no falsifiable prediction, because we know that supersymmetry, if it truly describes the world we see, must be broken. In chapter 5, we noted that supersymmetry may be seen in the LHC. This is possible but by no means guaranteed, even if supersymmetry is true.

Fortunately, there are other ways to test for supersymmetry. One possibility involves the dark matter. In many supersymmetric extensions of the standard model, the lightest new particle is stable and uncharged. This new stable particle could be the dark matter. It would interact with ordinary matter but only through gravity and the weak nuclear force. Such particles are called WIMPs, for weakly interacting massive particles, and several experiments have been mounted to detect them. These detectors exploit the idea that dark-matter particles will interact with ordinary matter via the weak force. This makes them very much like heavy versions of neutrinos, which also interact with matter only through gravity and the weak force.

Unfortunately, because the supersymmetric theories have so many free parameters, there is no specific prediction for what the mass of the WIMPs should be or exactly how strongly they should interact. But if they do indeed make up the dark matter, we can deduce what range is possible for their masses, assuming that they have played the role we think they have in the formation of galaxies. The range predicted is comfortably within the one that theory and experiment suggest for the lightest superpartner.

Experimentalists have looked for WIMPs by using detectors similar to those used to detect neutrinos coming from the sun and distant supernovas. Extensive searches have been carried out, but so far no WIMPs have been found. This is of course not definitive — it means only that if they exist, they interact too weakly to have triggered a response from a detector. What can be said is that if they interacted as strongly with matter as neutrinos do, they would have been seen by now. Still, the discovery of supersymmetry by any means would be a spectacular triumph for physics.

The main thing to keep in mind is that even if string theory requires that the world be supersymmetric on some scale, it makes no prediction about what that scale is. Thus, if supersymmetry is not seen in the LHC, that does not falsify string theory, because the scale at which it might be seen is completely adjustable. On the other hand, if supersymmetry is seen, that does not confirm string theory. There are ordinary theories that require supersymmetry, such as the minimal supersymmetric extension of the standard model. Even among quantum theories of gravity, supersymmetry is not unique to string theory; for example, the alternative approach called loop quantum gravity is completely compatible with supersymmetry.

We now come to the third generic prediction of string theory: that all the fundamental forces become unified at some scale. As in the other cases, this idea is broader than string theory, so a confirmation of it would not prove that string theory is right; indeed, string theory allows several possible forms of unification. But there is one form that most theorists believe represents *grand unification.* As we discussed in chapter 3, grand unification makes a generic prediction, so far unverified, that protons should be unstable and decay at some time scale. Experiments have looked for proton decay and failed to find it. These results (or lack of them) killed certain grand unified theories but not the general idea. However, the failure to find proton decay remains a constraint on possible theories, including supersymmetric theories.

A large number of theorists believe all three of these generic predictions will be confirmed. Consequently, experimentalists have put enormous effort into searching for evidence that would support them. It is not an exaggeration to say that hundreds of careers and hundreds of millions of dollars have been spent in the last thirty years in the search for signs of grand unification, supersymmetry, and higher dimensions. Despite these efforts, no evidence for any of these hypotheses has turned up. A confirmation of any of these ideas, even if it could not be taken as a direct confirmation of string theory, would be the first indication that at least some part of the package deal that string theory requires has taken us closer to, rather than further from, reality.

12

What String Theory Explains

WHAT ARE WE to make of the strange story of string theory thus far? More than two decades have now passed since the first superstring revolution. During this time, string theory has dominated the attention and resources of theoretical physics worldwide — more than a thousand of the world's most talented and highly trained scientists have worked on it. While there has been room for honest disagreement about the theory's prospects, sooner or later science is supposed to accumulate evidence that allows us to reach a consensus about the truth of a theory. Mindful that the future is always open, I would like to close this section by offering an assessment of string theory *as a proposal for a scientific theory.*

Let me be clear. First, I am not assessing the quality of the work; many string theorists are brilliant and well trained and their work is of the highest quality. Second, I want to separate the question of whether string theory is a convincing candidate for a physical theory from the question of whether or not research in the theory has led to useful insights for mathematics or other problems in physics. No one disputes that a lot of good mathematics has come out of string theory and that our understanding of some gauge theories has been deepened. But the usefulness of spin-offs for mathematics or other areas of physics is not evidence either for or against the correctness of string theory as a scientific theory.

What I want to assess is the extent to which string theory has fulfilled its original promise as a theory that unites quantum theory, gravity, and elementary-particle physics. String theory either is or is not the culmination of the scientific revolution that Einstein began in 1905. This kind of assessment cannot be based on unrealized hypotheses or unproved conjectures, or on the hopes of the theory's adherents. This is science, and the truth of a theory can be assessed based only on results that have been published in the scientific literature; thus we must be careful to distinguish between conjecture, evidence, and proof.

One might ask whether it is too early to make such an assessment. But string theory has been under continuous development for more than thirty-five years, and for more than twenty it has captured the attention of many of the brightest scientists in the world. As I emphasized earlier, there is no precedent in the history of science, since at least the late eighteenth century, for a proposed major theory going more than a decade before either failing or accumulating impressive experimental and theoretical support. Nor is it convincing to point to the experimental difficulties, for two reasons: First, much of the data that string theory was invented to explain already exists, in the values of the constants in the standard models of particle physics and cosmology. Second, while it is true that strings are too small to observe directly, previous theories have almost always quickly led to the invention of new experiments — experiments that no one would have thought of doing otherwise.

In addition, we have a lot of evidence to consider in making our evaluation. The many people working on string theory have given us a great deal to work with. Equally informative are the conjectures and hypotheses that have remained open despite intensive investigation. Most of the key conjectures that are unresolved are at least ten years old, and there is no sign that they will be resolved soon.

Finally, string theory is, as a result of the discovery of the vast landscape of theories described in chapter 10, in a crisis that is leading many scientists to reconsider its promise. So, while we must remember that new developments could change the picture, this appears to be a good time to attempt an assessment of string theory as a scientific theory.

The first step in the assessment of any theory is a comparison with observation and experiment. This was discussed in the last chapter. We learned that even after all the work that has been put into string theory, there is no realistic possibility for a definitive confirmation or falsification of a unique prediction from it by a currently doable experiment.

Some scientists would take that as reason enough to give up, but string theory was invented to solve certain theoretical puzzles. Even absent experimental test, we might be willing to support a theory that provided convincing solutions to outstanding problems. In the first chapter, I described the five major problems facing theoretical physics. The theory that will close Einstein's revolution should resolve all of them. It is thus fair to assess string theory by asking how well it does this.

Let's begin by recapping exactly what we know about string theory.

There is, first of all, no complete formulation of it. There is no accepted proposal for what the basic principles of string theory are, or for what the main equations of the theory should be. Nor is there proof that such a complete formulation exists. What we know of string theory consists mostly of approximate results and conjectures that concern the following four classes of theories.

1. The best-understood theories feature strings moving in simple backgrounds, such as flat ten-dimensional spacetime, where the geometry of the background is unchanging in time and the cosmological constant is zero. There are also many cases where some of the nine spatial dimensions are curled up, while the rest remain flat. These are the theories we understand best, because detailed calculations can be done of strings and branes moving and interacting in these backgrounds.

   In these theories, we describe the motion and interaction of the strings on the background spaces in terms of an approximation procedure called *perturbation theory*. What has been proved is that these theories are well defined and give finite and consistent predictions up to a second order in that approximation scheme. Other results support, but so far do not prove, the consistency of these theories. In addition, a large number of results and conjectures describe a network of duality relations among these theories.

However, every one of these theories disagrees with established facts about our world. Most of them have unbroken supersymmetry, which is not observed in the real world. The few that don't have unbroken supersymmetry predict that fermions and bosons have superpartners of equal mass, which is also not observed, and they also predict the existence of infinite-range forces in addition to gravity and electromagnetism, which again are unobserved.

2. In the case of a world with a negative cosmological constant, there is an argument for the existence of a class of string theories based on the Maldacena conjecture. This relates string theory on certain spaces with a negative cosmological constant to certain supersymmetric gauge theories. So far, these string theories cannot be explicitly constructed and studied except for certain very special, highly symmetric extremal cases. The weaker versions of the Maldacena conjecture are supported by a lot of evidence, but it is unknown exactly which version of the conjecture is true. If the strongest version is true, then string theory is equivalent to gauge theory, and this relation provides an exact description of string theories with a negative cosmological constant. However, these theories also cannot describe our universe, because we know the cosmological constant is positive.

3. An infinite number of other theories are conjectured to exist that correspond to strings moving on more complicated backgrounds, in which the cosmological constant is not zero, in which the spacetime background geometry is evolving in time, or in which the background involves branes and other fields. This includes a vast number of cases where the cosmological constant is positive, in agreement with observation. It has so far been impossible to precisely define these string theories or do explicit calculations to draw predictions from them. The evidence for their existence is based on the satisfaction of certain necessary but far from sufficient conditions.

4. In twenty-six spacetime dimensions, there is a theory, without

fermions or supersymmetry, called the bosonic string. This theory has tachyons, which lead to infinite expressions, rendering the theory inconsistent.

It has been suggested that all the conjectured and constructed theories are unified in a deeper theory, called M-theory. The basic idea is that all the theories we understand will correspond to solutions of this deeper theory. There is evidence for its existence in the many duality relationships that are conjectured or demonstrated to hold among the various string theories, but so far no one has been able to formulate its basic principles or write down its basic laws.

From this summary, we can see why any evaluation of string theory will necessarily be controversial. If we restrict our attention to the theories that are known to exist — those that allow us to do actual calculations and make predictions — we must conclude that string theory has nothing to do with nature, because every single one of these disagrees with experimental data. So the hope that string theory may describe our world rests wholly on a belief in string theories whose existence is only conjectured.

Nevertheless, many working string theorists believe that the conjectured theories exist. This belief seems to be based on indirect reasoning, as follows:

1. They conjecture that a general formulation of string theory exists and is defined by unknown principles and unknown equations. This unknown theory is conjectured to have many solutions, each of which provides a consistent theory of strings propagating on some background spacetime.
2. They then write down equations that are conjectured to approximate the true equations of the unknown theory. It is then conjectured that these approximate equations give necessary but not sufficient conditions for a background to have consistent string theories. These equations are versions of Kaluza-Klein theory, in that they involve general relativity extended to higher dimensions.
3. For every such solution to these approximate equations, they conjecture the existence of a string theory, even if they cannot write it down explicitly.

The problem with this reasoning is that the first step is a conjecture. We do not know that the theory or the equations that would define it really exist. That makes the second step a conjecture as well. Also, we do not know that the conjectured approximate equations give us sufficient, as opposed to necessary, conditions for a string theory to exist.

There is a danger with this kind of reasoning — with assuming what needs to be proved. If you believe in the assumptions of the argument, then the theories whose existences are implied can be studied as examples of string theories. But it must be remembered that they are *not* string theories, nor theories of any kind, but rather solutions of classical equations. Their significance depends entirely on the existence of theories that no one has been able to formulate and conjectures that no one has been able to prove. Given this, there appear to be no compelling reasons to believe that any string theory that has not been explicitly constructed exists.

What conclusions can be drawn from all this? First, given the incomplete state of knowledge of string theory, there is a wide range of possible futures. Based on what we know now, a theory may well emerge that fulfills the original hopes. It is also possible that there is no real theory there and that all there ever will be is a large set of approximate results about special cases that hold only because they are constrained by special symmetries.

The inevitable conclusion seems to be that string theory itself — that is, the theory of strings moving on background spacetimes — is not going to be a fundamental theory. *If string theory is to be relevant at all for physics, it is because it provides evidence for the existence of a more fundamental theory.* This is generally recognized, and the fundamental theory has a name — M-theory — even if it has not yet been invented.

This may not be as bad as it seems. For example, most quantum field theories are not known to exist at a rigorous level. The quantum field theories that particle physicists study — including quantum electrodynamics, quantum chromodynamics, and the standard model — share with string theory the fact that they are defined only in terms of an approximation procedure. (Although these theories have been proven to give finite and consistent results to all orders of approximation.) Still, there is good reason to believe that the standard model does not exist as a rigorously defined mathematical theory. This is not disturbing, as long as we believe that the standard model is only a step toward a deeper theory.

String theory was at first thought to be that deeper theory. On the present evidence, we must admit that it is not. Like the quantum field theories, string theory seems to be an approximate construction that (to the extent it is relevant for nature) points to the existence of a more fundamental theory. This does not necessarily make string theory irrelevant, but to prove its worth it must do at least as well as the standard model. It must predict something new that turns out to be true and it must explain phenomena that have been observed. We have seen that so far it does not do the first. Does it do the second?

We can answer this by assessing how well string theory answers the five key problems outlined in chapter 1.

Let's start with the good news. String theory was originally motivated by the third problem, the problem of *unification of the particles and forces.* How has it stood up as such a unifying theory?

Quite well. On the backgrounds where consistent string theories are defined, the vibrations of a string include states that correspond to all the known kinds of matter and forces. The graviton, the particle that carries the gravitational force, comes out of the vibrations of loops (i.e., closed strings). The photon, carrier of the electromagnetic force, also emerges from the vibrations of a string. The more complicated gauge fields, in terms of our understanding of the strong and weak nuclear forces, also come out automatically; that is, string theory predicts generally that there are gauge fields similar to these, although it does not predict the particular mix of forces we see in nature.

Thus — at least on the level of the bosons, or force-carrying particles, on a background spacetime — string theory unifies gravity with the other forces. All four fundamental forces arise as vibrations of one fundamental kind of object, a string.

What about unifying the bosons with the particles that make up matter, like quarks and electrons and neutrinos? It turns out that these also arise as states of vibration of strings, when supersymmetry is added. Thus, supersymmetric string theories unify all the different kinds of particles with one another.

Moreover, string theory does all this with a simple law: that the strings propagate through spacetime so as to take up the least amount of area. Nor is there any need to have separate laws describing how particles interact; the laws by which strings interact follow directly from the simple law that describes how they propagate. And since the various forces and particles all are just vibrations of strings, the laws that describe them follow as well. Indeed, the whole set of equations describing the propagation and interactions of the forces and particles has been derived from the simple condition that a string propagates so as to take up the least area in spacetime. The beautiful simplicity of this is what excited us originally and what has kept many people so excited: a single kind of entity, satisfying a single simple law.

What about the first problem in chapter 1, the *problem of quantum gravity?* Here the situation is mixed. The good news is that the particles carrying the gravitational force come out of the vibrations of strings, as does the fact that the gravitational force exerted by a particle is proportional to its mass. Does this lead to a consistent unification of gravity with quantum theory? As I stressed in chapters 1 and 6, Einstein's general theory of relativity is a background-independent theory. This means that the whole geometry of space and time is dynamical; nothing is fixed. A quantum theory of gravity should also be background-independent. Space and time should arise from it, not serve as a backdrop for the actions of strings.

String theory is not currently formulated as a background-independent theory. This is its chief weakness as a candidate for a quantum theory of gravity. We understand string theory in terms of strings and other objects moving on fixed classical background geometries of space that don't evolve in time. So Einstein's discovery that the geometry of space and time is dynamical has not been incorporated into string theory.

It is interesting to reflect that apart from a few special one-dimensional theories, no rigorous background-dependent quantum field theory exists. All are defined only in terms of approximation procedures. Perhaps string theory shares this property because it is background-dependent. It is tempting to suggest that any consistent quantum field theory must be background-independent. If true, this would imply that unification of quantum theory with general relativity is not optional, it is forced.

There are claims that general relativity can, in a certain sense, be derived from string theory. This is a significant claim, and it is important to understand the sense in which it is true, for how can a background-independent theory be derived from a background-dependent theory? How can a theory in which the geometry of spacetime is dynamical be derived from a theory that requires a fixed geometry?

The argument for this is as follows: Consider a spacetime geometry and ask if there is a consistent quantum-mechanical description of strings moving and interacting in that geometry. When you investigate this proposition, you find that a necessary condition for the string theory to be consistent is that, to a certain approximation, the spacetime geometry is a solution to the equations of a higher-dimensional version of general relativity. So there is a sense in which the equations of general relativity *emerge* from the conditions for a string to move consistently. This is the basis for the claims that string theorists make for general relativity's derivation from string theory.

There's a catch, though. What I have just described is the situation in the original twenty-six-dimensional bosonic string. But, as noted, this theory has an instability, the tachyon, so it is not really a viable theory. To make the theory stable, one can make it supersymmetric. And supersymmetry gives rise to additional necessary conditions that the background geometry must satisfy. Currently the only supersymmetric string theories known in detail to be consistent live on background spacetimes that do not evolve in time.[1] So in these cases it cannot be asserted that all of general relativity is recovered as an approximation in supersymmetric string theory. It is true that many solutions to general relativity are recovered, including all the solutions in which some dimensions are flat and others are curled up. But these are very special; the generic solution of general relativity describes a world whose spacetime geometry changes in time. This captures Einstein's essential insight that the geometry of spacetime is dynamical and evolving. You cannot recover only those solutions with no time dependence and still say that general relativity is derived from string theory. Nor can you claim to have a theory of gravity, since many gravitational phenomena involving time dependence have been observed.

In response, some string theorists conjecture that there are consistent string theories on spacetime backgrounds that vary in time but these are just much more difficult to study. They cannot be supersymmetric, and to my knowledge, there is no explicit general construction of such theories. The evidence for them is of two kinds. First, there is an argument that at least small amounts of time dependence can be introduced without disturbing the conditions required to eliminate the tachyon and make the theory consistent. This argument is plausible, but with no detailed construction it's hard to judge. Second, some special cases have been worked out in detail; however, the most successful of these have a hidden symmetry in time, so they don't fit. Others have possible problems with instabilities, or are worked out only at the level of classical equations that don't go far enough to show whether or not they really exist. Still others have a very fast time dependence, governed by the scale of the string theory itself.

In the absence of an explicit construction of a string theory on a general time-dependent spacetime, or a compelling argument for its existence without assuming the existence of a meta-theory, it cannot now be asserted that all of general relativity can be derived from string theory. This is another issue that remains open, to be decided by future work.

One can still ask whether string theory gives a consistent theory that includes gravity and quantum theory in those cases where the theory can be constructed explicitly. That is, can we at least describe gravitational waves and forces so weak that they can be seen as barely rippling the geometry of space? And can we do this completely consistently with quantum theory?

This can be done to a certain approximation. So far, the attempts to prove it beyond that level of approximation have not fully succeeded, although a lot of positive evidence has been gathered and no counterexample has emerged. Certainly, it's widely believed by string theorists to be true. At the same time, the obstacles to proving it seem substantial. The approximation method, perturbation theory, gives answers to any physical question by a sum of an infinite number of terms. For the first several terms, each one is smaller than the one before, so you get an approximation just by calculating a few terms. This is what is usually done in string theory and quantum field theory. To prove the theory finite, then, you have to prove that for any calculation you might do to answer a physical question, each of the infinite number of terms is finite.

Here is where things stand now. The first term is obviously finite, but that corresponds to classical physics, so there is no quantum mechanics in it. The second term, the first that could possibly be infinite, can easily be shown to be finite. It took until 2001 for a complete proof of the finiteness of the third term. It was heroic work, carried out over many years by Eric D'Hoker at UCLA and his collaborator Duong H. Phong, at Columbia University.[2] They have since then been working on the fourth term. They understand a great deal about this term but have no proof so far that it is finite. Whether or not they succeed in proving all the infinite terms finite remains to be seen. Part of the problem they face is that the algorithm for writing down the theory becomes ambiguous past the second term, so they need to first find the right definition for the theory before they can try proving that it gives finite answers.

How can this be? Haven't I pointed out that string theory is based on a very simple law? The problem is that the law is simple only when applied to the original theory in twenty-six dimensions. When supersymmetry is added, it becomes quite a bit more complicated.

There are additional results, which show for every term that certain possible infinite expressions that might have occurred do not in fact appear. A powerful proof of this sort was published in 1992 by Stanley Mandelstam. Recently a great deal of progress has been made by Nathan Berkovits, an American physicist who chooses happily to work in São Paulo. Berkovits has invented a new formulation of superstring theory. He achieves a proof good for each term in the perturbation theory, subject only to a couple of additional assumptions. It is too early to tell if those additional assumptions will be easy to lift. Still, this is substantial progress toward a proof. The issue of finiteness is not one that gets much attention from most string theorists, and I have enormous respect for the few that are still working hard on the problem.

There is one more worrying issue surrounding finiteness. In the end, even if each term in a calculation turns out to be finite, the exact answers to the calculation are derived by summing up all the terms. Since there are an infinite number of terms to be added, the result could again be infinite. While this summing up has not yet been done, there is evidence (too technical to describe here) that the result will be infinite. Another way to say this is that the approximation procedure comes only so close to the real predictions and then diverges from them. This is a common feature of quantum theories. It means that perturbation theory, while a useful tool, cannot be used to define the theory.

On present evidence, without a proof or a counterexample in hand, it is simply impossible to know whether string theory is finite. The evidence can be read either way. After a lot of hard work (albeit by just a handful of people), there are several partial proofs. This can be read either as clear evidence that the conjecture is true or that something is amiss. If such gifted physicists have tried and failed, and if every attempt remains incomplete, it may be because the conjecture they are trying to prove is false. The reason that mathematics invented the idea of proof and made it the criterion for belief is that human intuition has so often proved faulty. Widely believed conjectures sometimes turn out to be false. This is not an issue of mathematical rigor. Physicists don't typically aspire to the same level of rigor demanded by their mathematician cousins. There are a number of interesting and widely accepted theoretical results that don't have mathematical proof. But this is not one of those cases. There is no proof of string theory's finiteness even at a physicist's level of rigor.

Given this, I don't have a view on whether supersymmetric string theory will turn out to be finite or not. But if something so central to the claims of a theory is thought to be true, work should be expended to turn that intuition into proof. To be sure, there are many cases where a popular conjecture remains unproved for generations, but this is usually because key insights are missing. Even if the end result proves what everyone believes anyway, the effort is usually repaid by our gaining a much deeper insight into the area of mathematics that gave birth to the conjecture in the first place.

We will come back to why the finiteness of string theory is such a controversial issue. For now, we should simply note that it's not an isolated example. Several of the key conjectures that drove the two string revolutions remain unproved. These include the strong-weak duality and the Maldacena duality. In both cases, there is a lot of evidence that some form of relation between different theories is true. Even if the strict equivalences claimed in the conjectures are false, these are important ideas and results. But in any hard-nosed accounting, we have to distinguish between conjecture, evidence, and proof.

Some claim that the Maldacena conjecture offers independent proof that string theory yields a good quantum theory of gravity, at least in the case of certain geometries. They assert that string theory is in some cases precisely equivalent to an ordinary gauge theory in three spatial dimensions, giving a good quantum theory of gravity reliable to any order of approximation.

The problem with this assertion is that, as noted, the strong form of the Maldacena conjecture remains unproved. There is impressive evidence for some relationship between Maldacena's ten-dimensional supersymmetric string theory and the maximally super gauge theory, but what we currently have is not yet a proof of the full conjecture. The evidence is just as easily explained by there being only a partial correspondence between the two theories, neither of which is precisely defined. (Recently there has been progress on approaching the gauge theory through a second approximation scheme, called *lattice gauge theory*.) The present evidence is consistent with Maldacena's conjecture of complete equivalence being false, either because the two theories are in fact different or because one or both of them do not, strictly speaking, exist. On the other hand, if the strong form of the Maldacena conjecture turns out to be true — which is also consistent with the present evidence — then string theory provides good quantum theories of gravity, in the special case of backgrounds with a negative cosmological constant. Moreover, those theories would be partly background-independent, in that a nine-dimensional space is generated from physics in a three-dimensional space.

There is other evidence that string theory can provide a unification of gravity with quantum theory. The strongest results involve branes and black holes. These results are extraordinary, but, as we discussed in chapter 9, they do not go far enough. They are so far limited to very special black holes, and there appears to be little hope that the exact results will soon be extended to black holes in general, including the kinds believed to exist in nature; the results may be due to the extra symmetry these special black holes have. Finally, the string theory results do not include an actual description of the quantum geometry of the special black holes; they are limited to the study of model brane systems that share many properties with black holes but exist in an ordinary flat spacetime, and they are studied in an approximation in which the gravitational force is turned off.

Some argue that these extremal brane systems will become black holes when the gravitational force is turned back on. But string theory cannot follow up that argument with a detailed description of how a black hole is formed. To do this, you would need a string theory that worked in a spacetime that was evolving in time, and we have seen that this does not now exist.

Since the original results on black holes, there have been a number of imaginative proposals for how to describe real black holes in string theory. But they all suffer from a general problem, which is that whenever they stray from the very special black holes where we can use supersymmetry to do calculations, they fail to lead to precise results. Once we study ordinary black holes, or when we try to go inside to ask what happens to the singularity, we are unavoidably in the regime where the spacetime geometry evolves in time. Supersymmetry cannot work here, and neither do all the beautiful calculational tools that depend on it. So we are left with the same dilemma that afflicts so much research in string theory: We get marvelous results for very special cases, and we are unable to decide whether the results extend to the whole theory or are true only of the special cases where we can do the calculations.

Given these limitations, can it be claimed that string theory resolves the puzzles of black-hole entropy, temperature, and information loss implied by the discoveries of Jacob Bekenstein and Stephen Hawking? The answer is that while there are suggestive results, string theory cannot yet claim to have solved these problems. For the extremal and near-extremal black holes, calculations using the model systems of branes do reproduce all the details of the formulas that describe the thermodynamics of the corresponding black holes. But these are not black holes, they are just systems constrained by the requirements of having a large amount of supersymmetry to have the thermal properties of black holes. The results do not provide a description of the actual quantum geometry of black holes. Hence they do not explain Bekenstein and Hawking's results in terms of a microscopic description of black holes. Moreover, as noted, the results apply only to a very special class of black holes and not to those of real physical interest.

To summarize: On the basis of current results, we cannot confidently assert that string theory solves the problem of quantum gravity. The evidence is mixed. To a certain approximation, string theory seems to consistently unify quantum theory and gravity and give sensible and finite answers. But it is hard to decide if this is true of the whole theory. There is evidence to support something like the Maldacena conjecture, but no proof of the full conjecture itself, and only the full conjecture will allow us to assert the existence of a good quantum theory of gravity. The black-hole picture is impressive but only for the atypical black holes that string theory is able to model. Beyond these, there is the ever-present problem that string theory is not background-independent and even within that limitation cannot so far describe anything other than static backgrounds where the geometry does not evolve in time.

What we *can* say is that within these limitations there is some evidence that string theory points to the existence of a consistent unification of gravity and quantum theory. But is string theory itself that consistent unification? In the absence of a solution to these problems, it seems unlikely.

Let's turn to the other problems on chapter 1's list. The fourth problem is to explain the values of the parameters of the standard model of particle physics. It is clear that string theory has failed so far to do this, and there is no reason to believe it can. Instead, as we discussed in chapter 10, the evidence suggests that there are such a vast number of consistent string theories that the theory will make few if any predictions on this point.

The fifth problem is to explain what the dark matter and dark energy are, and to explain the values of the constants in cosmology. Here the situation is also not good. String theories, since they typically include many more particles and forces than have been observed, do offer a number of candidates for the dark matter and energy. Some of the extra particles could be the dark matter. Some of the extra forces could be the dark energy. But string theory offers no *specific* predictions as to which of the many possible candidates are the dark matter or the dark energy.

For example, among the possible dark-matter candidates is a particle called an *axion* (the word refers to certain properties, which I won't go into). Many, but not all, string theories contain axions, so at first this seems good. But most string theories that contain axions predict that they have properties that disagree with the standard cosmological model. So this seems bad. But then there are so many string theories that some may well contain axions consistent with the cosmological model. It is also possible that the cosmological model is wrong on this point. So it is reasonable to say that if axions are the dark matter, this is consistent with string theory. But this is very far from saying that string theory either predicts that the dark matter is an axion or makes any additional predictions by which observations of the dark matter could falsify string theory.

The remaining problem on our list is no. 2: *the foundational problems of quantum mechanics.* Does string theory offer any solution to those problems? The answer is no. String theory so far says nothing directly about the problems in the foundations of quantum theory.

So here is the accounting. Out of the five key problems, string theory potentially solves one of them completely, the problem of the unification of particles and forces. This is the problem that motivated the invention of string theory and it is still its most impressive success.

There is evidence that string theory points to a solution to the problem of quantum gravity, but at best it points to the existence of a deeper theory that solves the problem of quantum gravity rather than being itself the solution.

At the present time, string theory does not solve any of the three remaining problems. It appears incapable of explaining the parameters of the standard models of physics and cosmology. It provides a list of possible candidates for the dark matter and energy but doesn't uniquely predict or explain anything about them. And so far, string theory has nothing to say about the greatest mystery of all, which is the meaning of quantum theory.

Beyond this, are there any successes to speak of? One place we usually look for successes of a theory is in the predictions it makes for new experiments or observations. As we have said, string theory makes absolutely no predictions of this kind. Its strength is that it unifies the kinds of particles and forces we know about. If we didn't know about gravity, for example, we could predict its existence from string theory. This is not nothing. But it is not a prediction for a new experiment. Furthermore, there is no possibility of falsifying the theory — proving it wrong — by finding that an experiment or observation disagrees with the predictions of the theory.

If string theory makes no new predictions, then we should at least ask how well it accounts for the data we have. Here the situation is peculiar. Because of the incomplete state of our knowledge, we have to divide the many possible string theories into two groups and investigate each one separately. The first group is made up of those string theories known to exist, and the second contains those that are conjectured to exist but not yet constructed.

Because of the recent observations that the expansion of the universe is accelerating, we're forced to focus on string theories in the second class, since they are the only ones that agree with these findings. But we do not know how to compute the probabilities for strings to move and interact in these theories. Nor are we able to show that these theories exist; the evidence we have for them is in their backgrounds satisfying certain necessary but far from sufficient conditions. So even in the very best case, if there is any string theory that describes our universe, new techniques will have to be invented to calculate predictions for experiments that work in these new theories. The known string theories, as noted, all disagree with observed facts about our world: Most have unbroken supersymmetry; the others predict that fermions and bosons come in pairs of equal mass; and they all predict the existence of new (and so far unobserved) infinite-range forces. It is hard to avoid the conclusion that, however well motivated, string theory has failed to realize the hopes so many held for it twenty years ago.

In the heyday of 1985, one of the most enthusiastic proponents of the new revolutionary theory was Daniel Friedan, then at the University of Chicago's Enrico Fermi Institute. Here is what he had to say in a recent paper:

> String theory failed as a theory of physics because of the existence of a manifold of possible background spacetimes. . . . The long-standing crisis of string theory is its complete failure to explain or predict any large distance physics. String theory cannot say anything definite about large distance physics. String theory is incapable of determining the dimension, geometry, particle spectrum and coupling constants of macroscopic spacetime. String theory cannot give any definite explanations of existing knowledge of the real world and cannot make any definite predictions. The reliability of string theory cannot be evaluated, much less established. String theory has no credibility as a candidate theory of physics.[3]

Still, many string theorists soldier on. But how is it that, in the face of the problems we have been discussing, so many bright people continue to work on string theory?

One point that string theorists are passionate about is that the theory is beautiful, or "elegant." This is something of an aesthetic judgment that people may disagree about, so I'm not sure how it should be evaluated. In any case, it has no role in an objective assessment of the accomplishments of the theory. As we saw in Part I, lots of beautiful theories have turned out to have nothing to do with nature.

Some young string theorists argue that even if string theory does not succeed as the final unification, it has spin-offs that aid our understanding of other theories. They particularly refer to the Maldacena conjecture, discussed in chapter 9, which provides a way to study certain gauge theories from calculations that are easy to do in the corresponding gravity theory. This certainly works well for theories with supersymmetry, but if it is to be relevant for the standard model, it must work well for the gauge theories that have no supersymmetry. In this case there are other techniques, and the question is how well the Maldacena conjecture compares with those. The jury is still out. A good test case is a simplified version of a gauge theory in which there are only two dimensions of space. It has recently been solved, using a technique that owes nothing to supersymmetry or string theory.[4] It can also be studied through a third approach — brute-force calculation by computer. The computer calculations are believed to be reliable, hence they can serve as a benchmark against which to compare the predictions of other approaches. Such a comparison shows that the Maldacena conjecture does not do as well as the other technique.*

Some theorists also point to potential advances in mathematics as a reason to continue working on strings. One such potential advance involves the geometry of the six-dimensional spaces that string theorists study as possible examples of the compactified dimensions. In some cases, unexpected and striking properties of these six-dimensional geometries have been predicted by using the mathematics of string theory. This is welcome, but we should be clear about what happened. There was no contact with physics. What happened took place purely on a mathematical plane: String theory suggested conjectures that relate different mathematical structures. The string theories suggested that properties of the six-dimensional geometries could be expressed as simpler mathematical structures that could be defined on the two-dimensional surfaces the strings sweep out in time. The name of such a structure is a *conformal field.* What was suggested is that properties of certain six-dimensional spaces were mirrored in the structures of these conformal field theories. This led to surprising relations between pairs of six-dimensional spaces. This is a wonderful spin-off from string theory. And for it to be useful, we do not have to believe that string theory is a theory of nature. For one thing, conformal field theory plays a role in many different applications, including condensed-matter physics and loop quantum gravity. So it has nothing uniquely to do with string theory.

There are other cases in which string theory has led to discoveries in mathematics. In one very beautiful case, a certain toy model of string theory called *topological string theory* has led to striking new insights into the topology of higher-dimensional spaces. However, this is not by itself evidence that string theory is true about nature, for the topological string theories are simplified versions of

---

* Very recently these new techniques have also been successfully applied to QCD in the real-world case of three spatial dimensions.

string theory and do not unify the forces and particles observed in nature.

More generally, the fact that a physical theory inspires developments in mathematics cannot be used as an argument for the truth of the theory as a physical theory. Wrong theories have inspired many developments in mathematics. Ptolemy's theory of the epicycles might well have inspired developments in trigonometry and number theory, but that does not make it right. Newtonian physics inspired the development of major parts of mathematics, and it continues to do so, but that did not save Newtonian physics when it disagreed with experiment. There are many examples of theories based on beautiful mathematics that never had any successes and were never believed, Kepler's first theory of the planetary orbits being the signal example. So the fact that some beautiful mathematical conjectures are inspired by a research program cannot save a theory that has no clearly articulated core principles and makes no physical predictions.

The troubles that string theory faces go straight to the roots of the whole enterprise of unification. In the first part of the book, we identified huge obstacles plaguing earlier unified theories — obstacles that led to their failure. Some of these involved attempts to unify the world by introducing higher dimensions. The geometry of the higher dimensions turned out to be far from unique and wracked with instabilities. The basic reason, as we saw in the earlier chapters, is that unification always has consequences, which imply the existence of new phenomena. In the good cases — such as Maxwell's theory of electromagnetism, the electroweak theory of Weinberg and Salam, special relativity, and general relativity — these new phenomena were quickly seen. These are the rare cases in which we can celebrate the unification. In other attempts at unification, the new phenomena are not quickly seen, or already disagree with observation. Rather than celebrating the consequences of unification, the theorist must cleverly try to hide the consequences. I know of no cases where this concealing of consequences led in the end to a good theory; sooner or later, the attempted unification was abandoned.

Both supersymmetry and higher dimensions have turned out to be cases in which enormous effort must be expended to hide the consequences of the proposed unifications. No two known particles turn out to be related by supersymmetry; instead, each known particle has an unknown partner, and you have to tune the many free parameters of such theories to keep the unknown particles from being seen. In the case of higher dimensions, almost all solutions of the theory disagree with observations. The rare solutions that do reveal something like our world are unstable islands in a vast sea of possibilities, nearly all of which look totally alien.[5]

Can string theory avoid the problems that befell the earlier higher-dimensional and supersymmetric theories? It's unlikely, if only because there is much more to hide than there was in either Kaluza-Klein theory or supersymmetric theories. The mechanism proposed by the Stanford group to stabilize the higher dimensions might work. But the cost is high, as it leads to a vast expansion of the landscape of conjectured solutions. Hence, the price of avoiding the problems that doomed Kaluza-Klein theory is, at best, to adopt a point of view that string theorists initially rejected, which is that a vast number of possible string theories have to be taken equally seriously as potential descriptions of nature. This means that the original hope for a unique unification, and hence for falsifiable predictions about elementary-particle physics, has to be given up.

In chapter 11, we discussed the claims by Susskind, Weinberg, and others that the landscape of string theories may be the road ahead for physics and found these claims unconvincing. Where, then, does that leave us? In a recent interview, Susskind claims that the stakes are to accept the landscape and the dilution in the scientific method it implies or give up science altogether and accept intelligent design (ID) as the explanation for the choices of parameters of the standard model:

> If, for some unforeseen reason, the landscape turns out to be inconsistent — maybe for mathematical reasons, or because it disagrees with observation — I am pretty sure that physicists will go on searching for natural explanations of the world. But I have to say that if that happens, as things stand now we will be in a very awkward position. Without any explanation of nature's fine-tunings we will be hard pressed to answer the ID critics. One might argue that the hope that a mathematically unique solution will emerge is as faith-based as ID.[6]

But this is a false choice. As we will see shortly, there are other theories that offer genuine answers to the five great questions, and they are progressing quickly. To give up string theory does not mean to give up science; it just means giving up one direction that was once favored but failed to deliver what was hoped for it, in order to focus attention on other directions that now seem more likely to succeed.

String theory succeeds at enough things so that it is reasonable to hope that parts of it, or perhaps something like it, might comprise some future theory. But there is also compelling evidence that something has gone wrong. It has been clear since the 1930s that a quantum theory of gravity must be background-independent, but there is still little progress toward such a formulation of string theory that could describe nature. Meanwhile, the search for a single, unique, unified theory of nature has led to the conjecture of an infinite number of theories, none of which can be written down in any detail. And if consistent, they lead to an infinite number of possible universes. On top of this, all the versions we can study in any detail disagree with observation. Despite a number of tantalizing conjectures, there is no evidence that string theory can solve several of the big problems in theoretical physics. Those who believe the conjectures find themselves in a very different intellectual universe from those who insist on believing only what the actual evidence supports. The very fact that such a vast divergence of views persists in a legitimate field of science is in itself an indication that something is badly amiss.

So, is string theory still worth studying, or should it be declared a failure, as some suggest? The fact that many hopes have been disappointed and many key conjectures remain unproved may well be reason enough for some to give up working on string theory. But they are not reasons for research to stop altogether.

What if sometime in the future someone found a way to formulate a string theory that led uniquely to the standard model of particle physics, was background-independent, and lived only in the three-dimensional nonsupersymmetric world we observe. Even if the prospects for finding such a theory seem slim, it is a possibility — underscoring the general wisdom that a diversity of research programs is healthy for science, a point we'll return to later.

So string theory is certainly among the directions that deserve more investigation. But should it continue to be regarded as the dominant paradigm of theoretical physics? Should most of the resources aimed at the solution of the key problems in theoretical physics continue to support research in string theory? Should other approaches continue to be starved in favor of string theory? Should only string theorists be eligible for the most prestigious jobs and research fellowships, as is now the case? I think the answer to all these questions must be no. String theory has not been successful enough on any level to justify putting nearly all our eggs in its basket.

What if there are no other approaches worth working on? Some string theorists have advocated supporting string theory because it is "the only game in town." I would argue that even if this were the case, we should strongly encourage physicists and mathematicians to explore alternative approaches. If there are no other ideas, well, let's invent some. Since there is no near-term hope that string theory will make falsifiable predictions, there is no particular hurry. Let's encourage people to look for a quicker path to answering the five key questions of theoretical physics.

In fact there *are* other approaches — other theories and research programs that aim to solve those five problems. And while most theorists have been focusing on string theory, a few people have been making a lot of progress on the development of these other areas. More important, there are hints of new experimental discoveries, not anticipated by string theory, that, if confirmed, will point physics in new directions. These new theoretical and experimental developments are the subject of the next part of the book.

III

BEYOND STRING THEORY

13

Surprises from the Real World

THE GREEK PHILOSOPHER HERACLITUS left us a lovely epigram: Nature loves to hide. It is so often true. There is no way Heraclitus could have seen an atom. No matter how much his fellow philosophers speculated about them, to see an atom was beyond any technology they might have imagined. These days, theorists make great use of nature's tendency to inscrutability. If nature is indeed supersymmetric, or has more than three dimensions of space, she has hidden it well.

But sometimes the opposite is true. Sometimes the key things are right in front of us, there for the seeing. Hiding in plain sight from Heraclitus were easily perceivable facts we now take for granted, like the principle of inertia or the constant acceleration of falling objects. Galileo's observations of motion on Earth did not use the telescope or the mechanical clock. As far as I know, they could have been made in Heraclitus's time. He had only to ask the right questions.

So, while we bemoan how hard it is to test the ideas behind string theory, we ought to wonder what might be hiding in plain sight around us. In the history of science, there have been many instances of discoveries that surprised scientists because they were not anticipated by theory. Are there observations today that we theorists have not asked for, that no theory invites — observations that could move physics in an interesting direction? Is there a chance that such observations have already been made but ignored because, if confirmed, they would be inconvenient for our theorizing?

The answer to these questions is yes. There are several recent experimental results that point to new phenomena unsuspected by most string theorists and particle physicists. None is completely established. In a few cases, the results are reliable but the interpretations are disputed; in others, the results are too new and surprising to have been widely accepted.[1] But they are worth describing here, because if any of these hints turns into a true discovery, then there are important features of fundamental physics that are not predicted by any version of string theory and will be hard to reconcile with it. Other approaches will then become essential, not optional.

Let us start with the cosmological constant, thought to represent the dark energy accelerating the universe's expansion. As discussed in chapter 10, this energy was not anticipated by string theory, nor by most theories, and we have no idea what sets its value. Many people have thought hard about this for years, and we are more or less nowhere. I don't have an answer either, but I have a proposal for how we might find one. Let's stop trying to account for the cosmological constant's value in terms of known physics. If there is no way to account for a phenomenon on the basis of what we know, then maybe this is a sign that we need to look for something new. Perhaps the cosmological constant is a symptom of something else, in which case it might have other manifestations. How are we to look for them, or recognize them?

The answer will be simple, because universal phenomena are ultimately simple. Forces in physics are characterized by just a few numbers — for example, the distance over which a force travels and a charge to tell us how strong it is. What characterizes the cosmological constant is a scale, which is the distance scale over which it curves the universe. We can call this scale R. It is about 10 billion light-years, or 10^{27} centimeters.[2] What is weird about the cosmological constant is that its scale is huge compared with other scales in physics. Scale R is 10^{40} times the size of an atomic nucleus and 10^{60} times the Planck scale (which is about 10^{-20} times the size of a proton). So it's logical to wonder whether scale R might reflect some totally new physics. A good approach would be to look for phenomena that happen on the same vast scale.

Does anything else happen on the scale of the cosmological constant? Let's start with cosmology itself. The most precise cosmological observations we have are measurements of the *cosmic microwave background.* This is the radiation left over from the Big Bang, which comes to us from all directions of the sky. The radiation is purely thermal — that is, random. It has been cooling as the universe expands, and it is now at a temperature of 2.7 degrees Kelvin. The temperature is uniform across the sky to a high degree of precision, but at the level of a few parts in 100,000 there are fluctuations in it (see Fig. 13, top). The pattern of these fluctuations gives us important clues to the physics of the very early universe.

Over the last decades, the temperature fluctuations of the microwave background have been mapped by satellites, balloon-borne detectors, and ground-based detectors. One way to understand what these experiments measure is to think of the fluctuations as if they were sound waves in the early universe. It is, then, useful to ask how loud the fluctuations are at different wavelengths. The results give us a picture, such as that in Fig. 13, bottom, which tells us how much energy there is at the various wavelengths.

The picture is dominated by a large peak, followed by several smaller peaks. The discovery of these peaks is one of the triumphs of contemporary science. They are interpreted by cosmologists to indicate that the matter filling the early universe was resonant, much like the head of a drum or the body of a flute. The wavelength at which a musical instrument vibrates is proportional to its size, and the same is true in the universe. The wavelengths of the resonant modes tell us how big the universe was when it first became transparent: that is, when the initial hot plasma devolved, or "decoupled," into separate realms of matter and energy some three hundred thousand years after the Big Bang, at which time the microwave background became visible. These observations are extremely helpful in tying down the parameters of our cosmological models.

Another feature we see in the data is that there is very little energy in the largest wavelength. This may be just a statistical fluctu-

−200μK   200μK

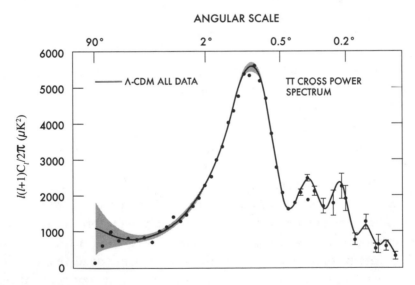

Fig. 13. Top: The sky as seen at microwave frequencies. Signals from within our galaxy have been removed, leaving an image of the universe as it was at the time when it cooled to the point that electrons and protons became bound into hydrogen. Bottom: The distribution of energy in the top image at different wavelengths. The dots represent data from WMAP and other observations, and the curve is a fit to predictions of the standard cosmological model.

ation, because it involves a small number of pieces of data. But if it is not a statistical fluke, it can be interpreted as indicating a cut-off, above which the modes are much less excited. It is interesting that this cutoff is at the scale R, associated with the cosmological constant.

The existence of such a cutoff would be puzzling from the point of view of the most widely accepted theory of the very early universe, which is *inflation*. According to the theory of inflation, the universe expanded exponentially fast during one extremely early period. Inflation accounts for the observation that the cosmic background radiation is so nearly uniform. It does this by ensuring that all the parts of the universe we see now could have been in causal contact when the universe was still a plasma.

The theory also predicts the fluctuations in the cosmic microwave background, which are hypothesized to be remnants of quantum effects during the period of inflation. The uncertainty principle implies that the fields dominating the energy of the universe during inflation fluctuate, and these fluctuations become imprinted on the geometry of space. As the universe expands exponentially, they persist, causing fluctuations in the temperature of the radiation produced when the universe becomes transparent.

Inflation is believed to have produced a huge region of the universe with relatively uniform properties. This region is thought to be many orders of magnitude larger than the observable region, because of a simple argument about scales. If inflation had stopped just at the point where it created a region as large as we now observe, there must have been some parameter in the physics of inflation that selected a special time to stop, which just happens to be our era. But this seems improbable, because inflation took place when the universe had a temperature ten to twenty orders of magnitude greater than the center of the hottest star today; thus the laws governing it must have been different laws, which dominated physics only in those extreme conditions. There are many hypotheses about the laws that governed inflation, and none of them say anything about a time scale of 10 billion years. Another way to put this is that there seems no way for the present value of the cosmological constant to have anything to do with the physics that caused inflation.

Thus if inflation produced a uniform universe on the scale that we observe, it likely produced a universe that is uniform on much larger scales. This in turn implies that the pattern of fluctuations produced by inflation should go on and on, no matter how far you look. If you could see beyond the present size of the observable universe, you should continue to see small fluctuations in the cosmic microwave background. Instead, the data hint that the fluctuations may cease above the scale R.

Indeed, as cosmologists have examined the large-scale modes in the microwave background, they have found more mysteries. It's an item of faith among cosmologists that at the largest scales the universe should be symmetric — that is, any one direction should be like any other. This is not what is seen. The radiation in these large-scale modes is not symmetric; there is a preferred direction. (It has been called the "axis of evil" by the cosmologists Kate Land and João Magueijo.[3]) No one has any rational explanation for this effect.

These observations are controversial because they disagree profoundly with what we would expect on the basis of inflation. Since inflation explains so much of cosmology, many prudent scientists suspect that there is something wrong with the microwave data. Indeed, it is always possible that the measurements are just wrong. A lot of delicate analysis is applied to the data before they're presented. One thing that's done is to subtract the radiation known to come from the galaxy we live in. This may have been done incorrectly, but few experts familiar with the details of how the data are analyzed believe that to be the case. Another possibility, as noted, is that our observations are just statistical anomalies. An oscillation at a wavelength of the scale R takes up a huge part of the sky — about 60 degrees; consequently we see only a few wavelengths, and there are only a few pieces of data, so what we are seeing may just be a random statistical fluctuation. The chances of the evidence for a preferred direction being a statistical anomaly have been estimated at less than 1 part in 1,000.[4] But it may be easier to believe in this unlikely bad luck than to believe that the predictions of inflation are breaking down.

These issues are currently unresolved. For the time being, it is enough to say that we went looking for strange physics on the scale R and found it.

Are there any other phenomena associated with this scale? We can combine R with other constants of nature to see what happens at scales defined by the resulting number. Let me give an example. Consider R divided by the speed of light: R/c. This gives us a time, and the time it gives us is roughly the present age of our universe. The inverse, c/R, gives us a frequency — a *very* low note, one oscillation per lifetime of the universe.

The next simplest thing to try is c^2/R. This turns out to be an acceleration. It is in fact the acceleration by which the rate of expansion of the universe is increasing — that is, the acceleration produced by the cosmological constant. Compared to ordinary scales, however, it is a very tiny acceleration: 10^{-8} centimeters per second. Imagine a bug crawling across the floor. It manages to go perhaps 10 centimeters per second. If the bug doubled its speed over the lifetime of a dog, it would be accelerating about as much as c^2/R, a very small acceleration indeed.

But suppose there is a new universal phenomenon that explains the value of the cosmological constant. Just by the fact that the scales match, this new phenomenon should also affect any other kind of motion with an acceleration this tiny. So anytime we can observe something moving with such a tiny acceleration, we would expect to see something new. Now the game starts to get interesting. We do know things that accelerate this slowly. One example is a typical star orbiting in a typical galaxy. A galaxy orbiting another galaxy accelerates even more slowly. So, do we see anything different about the orbits of the stars with accelerations this tiny, compared to the orbits of stars with larger accelerations? The answer is yes, we do, and dramatically so. This is the problem of the dark matter.

As we discussed in chapter 1, astronomers discovered the dark-matter problem by measuring the acceleration of stars in orbit about the center of their galaxies. The problem arose because, given the measured accelerations, astronomers could deduce the distribution of the galaxy's matter. In most galaxies, this result turned out to disagree with the matter observed directly.

I can now say a bit more about where the discrepancy arises. (For the sake of simplicity, I'll restrict the discussion to spiral galaxies, in which most stars move in circular orbits in a disk.) In each galaxy where the problem is found, it affects only stars moving outside a certain orbit. Within that orbit, there's no problem — the acceleration is what it should be if caused by the visible matter. So there seems to be a region in the interior of the galaxy within which Newton's laws work and there's no need for dark matter. Outside this region, things get messy.

The key question is: Where is the special orbit that separates the two regions? We might suppose that it occurs at a particular distance from the center of the galaxy. This is a natural hypothesis, but it is wrong. Is the dividing line at a certain density of stars or starlight? Again, the answer is no. What seems to determine the dividing line, surprisingly, is the rate of the acceleration itself. As one moves farther out from the center of the galaxy, accelerations decrease, and there turns out to be a critical rate that marks the breakdown of Newton's law of gravity. As long as the acceleration of the star exceeds this critical value, Newton's law seems to work and the acceleration predicted is the one seen. There is no need to posit any dark matter in these cases. But when the acceleration observed is smaller than the critical value, it no longer agrees with the prediction of Newton's law.

What is this special acceleration? It has been measured to be 1.2×10^{-8} centimeters per second per second. This is close to c^2/R , the value of the acceleration produced by the cosmological constant!

This remarkable twist in the dark-matter story was discovered by an Israeli physicist named Mordehai Milgrom in the early 1980s. He published his findings in 1983, but for many years they were largely ignored.[5] As the data have gotten better, however, it has become clear that his observation was correct. The scale c^2/R characterizes where Newton's law breaks down for galaxies. This is now called *Milgrom's law* by astronomers.

I want you to understand how weird this observation is. The scale R is the scale of the whole observable universe, which is enormously bigger than any individual galaxy. The acceleration c^2/R occurs on this cosmological scale; as noted, it is the rate at which the universe's expansion accelerates. There is no obvious reason for this scale to play any role at all in the dynamics of an individual galaxy. The realization that it does was forced on us by the data. I recall my amazement when I first learned about it. I was shocked and energized. I walked around for an hour in a daze, muttering incoherent obscenities. Finally! A possible hint from experiment that there is more to the world than we theorists imagine!

How is this to be explained? Apart from coincidence, there are three possibilities. There could be dark matter, and the scale c^2/R could characterize the physics of the dark-matter particles. Or the dark-matter halos could be characterized by the scale c^2/R, because that is related to the density of dark matter at the time they collapsed to form galaxies. In either case, the dark energy and dark matter are distinct phenomena, but related.

The other possibility is that there is no dark matter and Newton's law of gravity breaks down whenever accelerations get as small as the special value of c^2/R. In this case, there needs to be a new law that replaces Newton's law in these circumstances. In his 1983 paper, Milgrom proposed such a theory. He called it MOND, for "modified Newtonian dynamics." According to Newton's law of gravity, the acceleration of a body due to a mass decreases in a specific way when you move away from that mass — that is, by the square of the distance. Milgrom's theory says that Newton's law holds, but only until the acceleration decreases to the magic value of 1.2×10^{-8} cm/sec^2. After that point, rather than decreasing with the square of the distance, it decreases only by the distance. Moreover, while normally the Newtonian force is proportional to the mass of the body causing the acceleration times a constant (which is Newton's gravitational constant), MOND says that when the acceleration is very small, the force is proportional to the square root of the mass times Newton's constant.

If Milgrom is right, then the reason the stars outside the special orbit are accelerating more than they should be is that they are feeling a stronger gravitational force than Newton predicted! Here is brand-new physics — not at the Planck scale, and not even in an accelerator, but right in front of us, in the motions of the stars we see in the sky.

As a theory, MOND does not make much sense to physicists. There are good reasons why the gravitational and electrical forces fall off with the square of the distance. It turns out to be a consequence of relativity combined with the three-dimensional nature of space. I won't go into the details here, but the conclusion is drastic. Milgrom's theory appears inconsistent with basic physical principles, including those of special and general relativity.

There have been attempts to modify general relativity to construct a theory that contains MOND or something close to it. One such theory was invented by Jacob Bekenstein; another by John Moffat, then at the University of Toronto; and still another by Philip Mannheim, at the University of Connecticut. These are very imaginative people (Bekenstein, you'll recall from chapter 6, discovered black-hole entropy, while Moffat has invented many surprising things, including the variable-speed-of-light cosmology). All three theories work to some extent, but they are, to my mind, highly artificial. They have several extra fields and require the adjustment of several constants to unlikely values in order to get agreement with observation. I also worry about issues of instability, although the authors claim that such problems are settled. The good news is that people can study these theories the old-fashioned way — by comparing their predictions to the large amount of astronomical data we have.

It should be said that outside the galaxies, MOND does not work very well. There are a lot of data about the distribution of mass and the motion of galaxies on scales larger than the galactic scale. In this regime, the theory of dark matter does much better than MOND at accounting for the data.

Nevertheless, MOND seems to work quite well within galaxies.[6] Data gathered over the last decade have shown that in more than eighty cases (by last count) out of around a hundred studied, MOND predicts accurately how the stars are moving. In fact, MOND predicts how the stars move within galaxies *better* than the models based on dark matter. Of course, the latter are improving all the time, so I will not pretend to be able to predict how the match-up will turn out. But, for the present, we seem to face a delightfully frightful situation. We have two very different theories, only one of which can be right. One theory — the one based on dark matter — makes good sense, is easy to believe in, and does very well at predicting the motions outside galaxies but not so well inside them. The other theory, MOND, does very well with galaxies, fails outside the galaxies, and in any case is based on assumptions that seem to contradict extremely well-established science. I should confess that nothing has kept me awake at night in the past year more than worrying about this problem.

It would be easy to disregard MOND if not for the fact that Milgrom's law suggests that the scale of the mysterious cosmological constant somehow bears on whatever is determining how stars move in galaxies. Just from the data, it appears that the acceleration c^2/R plays a key role in how stars move. Whether this is because of a connection between dark matter and either dark energy or the cosmological expansion scale, or something more radical, we see that new physics can indeed be found at this acceleration.

I've had conversations about MOND with several of the most imaginative theorists I know. Often it went like this: We would be talking about some sober mainstream problem and one of us would mention galaxies. We would look at each other with a glint of recognition and one of us would say, "So you worry about MOND, too," as if admitting to a secret vice. Then we would share our crazy ideas — because all ideas about MOND that are not immediately wrong turn out to be crazy.

The only advantage is that this a case where there are lots of data, and the data get better all the time. Sooner or later we will know whether dark matter explains the motions of stars in galaxies, or we have to accept a radical modification of the laws of physics.

Of course, it could just be an accident that dark matter and dark energy share a common physical scale. Not all coincidences are meaningful. So we should ask whether there are other phenomena where this tiny acceleration can be measured. If so, are there situations where theory and experiment disagree?

It turns out that there *is* another such case, and it, too, is unsettling. NASA has by now sent several spacecraft out of the solar system. Of these, two — *Pioneers 10* and *11* — have been tracked for decades. The Pioneers were designed to tour the outer planets, after which they have continued moving away from the sun, in opposite directions in the plane of the solar system.

NASA scientists at the Jet Propulsion Laboratory (JPL) in Pasadena, California, are able to determine the velocities of the Pioneer spacecraft by using the Doppler shift and thus can accurately track their trajectories. JPL tries to anticipate the trajectories by predicting the forces on the two spacecraft from the sun, planets, and other constituents of the solar system. In both cases, the trajectories observed do not match those predicted.[7] The discrepancies are caused by an additional acceleration pulling the two spacecraft toward the sun. The size of this mysterious acceleration is around 8×10^{-8} centimeters per second per second — bigger than the anomalous acceleration value measured in galaxies by about a factor of 6. But it is still fairly close, given that there is no apparent connection between the two phenomena.

I should point out that the data in this case are not yet fully accepted. Whereas the anomaly is seen in both Pioneers, which is much more persuasive than if it were seen in only one, they were both built and are being tracked by JPL. However, the JPL data have been independently analyzed by scientists using Aerospace Corporation's Compact High Accuracy Satellite Motion Program, and these results agree with JPL's. So the data have stood up so far. But astronomers and physicists understandably have high standards of proof, especially when we're being asked to believe that Newton's law of gravity is breaking down just outside our solar system.

Since the discrepancy is small, it might possibly be accounted for by some tiny effect, such as the spacecraft side that faces the sun being slightly hotter than the opposite side, or by a small gas leak. The JPL team has taken into account every such effect they can think of and so far has been unable to explain the observed anomalous acceleration. Recently there have been proposals to send out a special-purpose probe, designed and built to eliminate as many such spurious effects as possible. Such a probe would take many years to leave the solar system, but even so, this is a worthwhile mission. Newton's law of gravity has stood for more than three hundred years; if it takes a few more either to confirm it or prove it wrong, that's not much to ask.

What if MOND or the Pioneer anomaly turn out to be correct? Might their data be reconciled with some existing theory?

At the very least, MOND is inconsistent with all the versions of string theory so far studied. Might it be consistent with some currently unknown version of string theory? Of course. Given string theory's flexibility, there is no way to rule this out, though it would be a difficult accomplishment. What about other theories? Several people have tried hard to make MOND come out of a brane-world scenario or some version of quantum gravity. There are a few ideas, but nothing that works impressively. Fotini Markopoulou, my colleague at the Perimeter Institute for Theoretical Physics, and I have speculated about how to get MOND from quantum gravity, but we haven't been able to show how our idea works in detail. MOND is a tantalizing mystery, but not one that can be resolved now, so let's move on to other hints of new physics coming from experiment.

The most dramatic experiments are those that overturn universally held beliefs. Some beliefs are so embedded in our thinking that they are reflected in our language. For example, we speak of the physical *constants*, to denote those numbers that never change. These include the most basic parameters of the laws of physics, such as the speed of light or the charge of the electron. But are these constants actually constant? Why couldn't the speed of light change with time? And could such a change be detected?

In the multiverse theory discussed in chapter 11, we imagined the parameters varying over a range of different universes. But how can we observe such variations in our own? Could the constants, such as the speed of light, change over time in our universe? Some physicists have pointed out that the speed of light is measured in some system of units — that is, so many kilometers per second. How, they argue, can you distinguish the speed of light varying over time in a situation in which the units themselves vary over time?

To answer this question, we have to know how the units of distance and time are defined. These units are based on some physical standard, which is defined in terms of the behavior of some physical system. At first, the standards referred to Earth: a meter was one-millionth the distance from the North Pole to the equator. Now the standards are based on properties of atoms — for example, a second is defined in terms of the vibrations of an atom of cesium.

If you take into account how the units are defined, then the physical constants are defined as ratios. For example, the speed of light can be defined if you know the ratio between the time it takes light to cross an atom and the period of light that the atom emits. These kinds of ratios are the same in all systems of units. The ratios refer purely to physical properties of atoms; no decision about choice of units is involved in measuring them. Since the ratios are defined in terms of physical properties alone, it is meaningful to ask whether these ratios change over time or not. If they do, then there is also a change over time in the relationship between one physical property of an atom and another.

Changes in these ratios would be measurable by changes in the frequencies of light emitted by atoms. Atoms emit light in a spectrum made up of many discrete frequencies, so there are many ratios defined by pairs of these frequencies. One can ask whether these ratios are different in the light from faraway stars and galaxies — that is, in light that is billions of years old.

Experiments of this kind have failed to detect changes in the constants of nature within our galaxy or among nearby galaxies. On time scales of millions of years, then, the constants have not changed in any detectable way. But an ongoing experiment by a group in Australia has found changes in the ratios by looking at light from quasars — light that was emitted on the order of 10 billion years ago. The Australian scientists don't study the atomic spectra of the quasar itself; what they do is more clever than that. On its way from the quasar to us, the light travels through many galaxies. Each time it passes through a galaxy, some light is absorbed by atoms in that galaxy. Atoms absorb light at specific frequencies, but because of the Doppler effect, the frequency at which the light is absorbed is shifted toward the red end of the spectrum by an amount proportional to that galaxy's distance from us. The result is that the spectrum of light from the quasar is decorated by a forest of lines, each corresponding to light being absorbed by a galaxy a particular distance from us. By studying ratios of the frequencies of these lines, we can look for changes in the fundamental constants over the time that the light has been traveling from the quasar. Because a change must show up as a frequency ratio and there are several fundamental constants, physicists have settled on the simplest ratio to study — the *fine structure constant*, which is made up of the constants that determine the properties of the atom. It is called *alpha*, and it is equal to the ratio of the charge of the electron squared, divided by the speed of light, times Planck's constant.

The Australians studied measurements of light from a sample of eighty quasars, using very precise spectra taken by the Keck telescope, in Hawaii. They deduced from their data that around 10 billion years ago, alpha was smaller by about 1 part in 10,000.[8]

This is a small change, but if it holds up, it is a momentous discovery, the most important in decades. This would be the first time that a fundamental constant of nature had been seen to vary over time.

Many of the astronomers I know are keeping an open mind. By all accounts, the data have been taken and analyzed with extreme care. No one has found an obvious flaw in the Australian team's method or results, but the experiment itself is very delicate, the precisions involved are at the edge of what is possible, and we cannot rule out the possibility that some error has crept into the analysis. As of this writing, the situation is messy, as is typical for a new experimental technique. Other groups are attempting the same measurements, and the results are controversial.[9]

Many theorists are skeptical of the indications of a variation in the fine structure constant. They worry that such a variation would be exceedingly unnatural, in that it would introduce to the theory of electrons, nuclei, and atoms a time scale many orders of magnitude removed from the scale of atomic physics. Of course, they could have said this about the scale of the cosmological constant. In fact, the scale at which the fine structure constant varies is not close to anything else that has been measured, except for the cosmological constant itself. So perhaps this is another mysterious phenomenon having to do with the scale R.

Yet another manifestation of the scale R may be the mysterious neutrino masses. You can convert the length scale R to a mass scale, using just the fundamental constants of physics, and the result is the same order of magnitude as the differences between the masses of the various kinds of neutrinos. No one knows why neutrinos, the lightest particles there are, should have masses related to R, but there it is — another tantalizing hint.

There could be a final experimental hint involving the scale R. By combining it with Newton's gravitational constant, we can conclude that there may be effects that alter the gravitational force on the scale of millimeters. Currently a group at the University of Washington, led by Eric Adelberger, has been making ultraprecise measurements of the force of gravity between two objects that are millimeters apart. As of June 2006, all they are able to say publicly is that they see no evidence that Newton's laws are wrong down to scales of $^6/_{100}$ of a millimeter.

If nothing else, our experiments should certainly test the fundamental principles of physics. There is a great tendency to think that these principles, once discovered, are eternal, yet history tells a different story. Almost every principle once proclaimed has been superseded. No matter how useful they are or how good an approximation they give to phenomena, sooner or later most principles fail, as experiment probes the natural world more accurately. Plato proclaimed that everything in the celestial sphere moves on circles. There were good reasons for this: Everything above the sphere of the moon was believed to be eternal and perfect, and no motion is more perfect than uniform motion in a circle. Ptolemy adopted this principle and enhanced it by constructing epicycles — circles moving on circles.

The orbits of the planets are indeed very nearly circular, and the motion of the planets in their orbits is almost uniform. Somehow it is fitting that the least circular of the planetary orbits is that of unruly Mars — and its orbit is so close to circular that the deviations are at the limit of what can be deduced from the best possible naked-eye observations. In 1609, after nine years of painstaking work on the Martian orbit, Johannes Kepler realized that it must be an ellipse. That year, Galileo turned a telescope to the sky and began a new era of astronomy, in which it eventually became clear that Kepler was right. Circles were the most perfect shapes, but planetary orbits are not circular.

When the ancients declared the circle the most perfect shape, they meant that it was the most symmetric: Each point on the orbit is the same as any other. The principles that are hardest to give up are those that appeal to our need for symmetry and elevate an observed symmetry to a necessity. Modern physics is based on a collection of symmetries, which are believed to enshrine the most basic principles. No less than the ancients, many modern theorists believe instinctively that the fundamental theory must be the most symmetric possible law. Should we trust this instinct, or should we listen to the lesson of history, which tells us that (as in the example of the planetary orbits) nature becomes less rather than more symmetric the closer we look?

The symmetries most deeply embedded in contemporary theory are those that come from Einstein's special and general theories of relativity. The most basic of these is the relativity of inertial frames. It is actually Galileo's principle, and it has been a foundational idea of physics since the seventeenth century. It says that we cannot distinguish motion with a constant speed and direction from rest. It is this principle that is responsible for the fact that we don't feel the motion of the earth, or our motion in an airplane cruising at constant speed in the sky. As long as there is no acceleration, you cannot feel your own motion. Another way to express this is that there is no preferred observer and no preferred frame of reference: As long as acceleration is absent, one observer is as good as another.

What Einstein did in 1905 was to apply this principle to light. A consequence is that the speed of light must be considered a constant, independent of the motion of the light source or the observer. No matter how we are moving relative to each other, you and I will attribute exactly the same speed to a photon. This is the basis of Einstein's special theory of relativity.

Given the special theory of relativity, we can make many predictions about the physics of the elementary particles. Here is one concerning cosmic rays. These are a population of particles, believed to be mainly protons, that travel through the universe. They arrive at the top of Earth's atmosphere, where they collide with atoms in the air, producing showers of other kinds of particles, which can be detected on the ground. No one knows the source of these cosmic rays, but the higher their energy, the rarer they are. They have been observed at energies more than 100 billion times the mass of the proton. To have this energy, protons must be moving very, very close to the speed of light — a speed limit that, according to special relativity, no particle is allowed to break.

Cosmic rays are believed to come from distant galaxies; if so, they must travel millions and perhaps billions of light-years across the universe before arriving here. Back in 1966, two Soviet physicists, Georgiy Zatsepin and Vadim Kuzmin, and (independently) the Cornell University physicist Kenneth Greisen made a striking prediction about cosmic rays, using only the special theory of relativity.[10] Their prediction, usually known as the GZK prediction, is worth describing, because it is just now being tested. It is the most extreme test of special relativity ever made. It is, in fact, *the first test of special relativity approaching the Planck scale, the scale at which we might see the effects of a quantum theory of gravity.*

Good scientists take advantage of all the tools at their disposal. What Greisen, Zatsepin, and Kuzmin understood is that we have access to a laboratory vastly larger than any we could build on Earth — the universe itself. We can detect cosmic rays that arrive on Earth after traveling for billions of years over a substantial proportion of the universe. As they travel, very small effects — effects that would be too tiny to show up in earthly experiments — may amplify to the point where we can see them. If we use the universe as an experimental tool, we can see much deeper into the structure of nature than people ever imagined.

The key point is that the space that cosmic rays are traveling through is not empty; it is filled with the cosmic microwave background radiation. Greisen and the Soviet scientists realized that protons of energy greater than a particular value would interact with the photons in the background radiation and that this interaction would create particles (most likely pions, or pi-mesons). This particle creation would take energy, and because energy is conserved, the high-energy protons would be slowed down. Thus, space is in effect opaque to the passage of any protons that carry more energy than that needed to make pions.

Space therefore functions as a kind of filter. The protons making up cosmic rays can travel only if they have less energy than that required to make pions. If they have more, they make pions and slow down and keep doing so until they slow to the point at which they can no longer make pions. It is as if the universe had imposed a speed limit on protons. Greisen, Zatsepin, and Kuzmin predicted that no protons would arrive at Earth with more than the energy needed to make pions in this way. The energy at which they predicted that this pion creation would happen is about a billionth of the Planck energy (10^{19} GeV) and is called the GZK cutoff.

This is an enormous energy, closer to the Planck energy than anything else we know of. It is more than 10 million times the energy that will be made in the most sophisticated particle accelerators currently planned. The GZK prediction provides a stringent test of Einstein's special theory of relativity. It probes the theory at a much higher energy, and for a velocity much closer to the speed of light, than any experiment done, or even feasible, on Earth. In 1966, when the GZK prediction was made, only cosmic rays with energies much lower than the predicted cutoff were being seen, but recently a few instruments have been built that can detect cosmic-ray particles at or even above the predicted cutoff. One such experiment, called AGASA (for Akeno Giant Air Shower Array), carried out in Japan, reported at least a dozen such extreme events. The energy involved in these events is greater than 3×10^{20} electron volts — roughly the energy a pitcher can give to a fastball, but all carried by one proton.

These events may be a signal that special relativity is breaking down at extreme energies. Sidney Coleman and Sheldon Glashow proposed in the late 1990s that a breakdown in special relativity could raise the energy required to make pions, thus raising the GZK cutoff energy and allowing protons of much higher energy to reach our detectors on Earth.[11]

This is not the only possible explanation for the observation of these higher-energy cosmic-ray protons. It is possible that they originated close enough to Earth so that they haven't had time to be slowed down by interaction with the cosmic microwave background. This can be checked by seeing if the protons in question arrived from any preferred place in the sky. There is so far no such evidence, but it remains a possibility.

It is also possible that these extremely high-energy particles are not protons at all. They could be a so-far-unknown species of stable particle with a mass much higher than that of protons. If so, this, too, would be a major discovery.

It is of course always possible that the experiments are wrong. The AGASA team report that their measurements of energy are accurate up to an uncertainty of about 25 percent, which is a big percentage of error, but still not enough to explain the existence of the highest-energy events they see. However, their estimate of their experiment's degree of accuracy could be wrong.

Luckily, an experiment now in progress will resolve the disagreement. This is the Auger cosmic-ray detector, now in operation on the pampas of western Argentina. If the Auger detectors confirm the Japanese observation and if the other possible explanations can be discounted, this would be the most momentous discovery of the last hundred years — the first breakdown of the basic theories comprising the twentieth century's scientific revolution.

What does it take to observe cosmic-ray particles of such an extreme energy? When a particle of this energy strikes the top of the atmosphere, it produces a shower of other particles that rain down over an area of many square kilometers. The Auger experiment consists of hundreds of detectors placed over 3,000 square kilometers of the Argentinean pampas. Several HiRes (for "high resolution") light sensors at the site also scan the sky to catch the light produced by the particle shower. By combining the signals made by all these detectors, the Auger researchers can determine the energy of the original particle that struck the atmosphere, as well as the direction from which it came.

As of this writing, the Auger Observatory is just releasing its first data. The good news is that the experiment is working well, but there is still not quite enough data to decide whether the cutoff predicted on the basis of special relativity is there or not. Still, it is reasonable to hope that after running for a few years, there will be enough data to settle the issue.

Even if the Auger team announces that special relativity remains viable, this finding alone will be the most important in fundamental physics in at least twenty-five years — that is, since the failure to find proton decay (see chapter 4). The long dark era in which theory developed without the guidance of experiment will finally be over. But if Auger discovers that special relatively is *not* completely right, it will usher in a new era in fundamental physics. It's worth taking some time to explore the implications of such a revolutionary finding and where it might lead.

14

Building on Einstein

SUPPOSE THAT THE AUGER project or some other experiment shows that Einstein's special theory of relativity breaks down. This would be bad news for string theory: It would mean that the first great experimental discovery of the twenty-first century was totally unanticipated by the most popular "theory of everything." String theory assumes that special relativity is true, exactly as written down by Einstein a hundred years ago. Indeed, a major achievement of string theory was to make a theory of strings consistent with both quantum theory and special relativity. So string theory predicts that no matter how distant their sources are from one another, photons of different frequencies travel at the same speed. As we have seen, string theory does not make many predictions, but this is one; in fact, it's the only prediction of string theory that can be tested by present technology.

What would it mean for the predictions of special relativity to be falsified? There are two possibilities. One is that special relativity is wrong, but the other possibility leads to a deepening of it. On this distinction rides a tale of perhaps the most surprising new idea to have emerged in fundamental physics in the last decade.

There are several experiments that could reveal a breakdown or modification of special relativity. The Auger experiment could do it, but so could our observations of gamma-ray bursts. These are enormous explosions that for a few seconds can produce as much light as that emitted by a whole galaxy. As the name implies, most of this light is radiated in gamma rays, which are a highly energetic form of photons. Signals from these explosions reach Earth on average about once a day. They were first detected in the late 1960s, by military satellites designed to look for illegal tests of nuclear weapons. Now they are observed by scientific satellites, whose purpose is to detect them.

We do not know exactly what the sources of gamma-ray bursts are, although there are plausible theories. They may come from the collision of two neutron stars or of a neutron star and a black hole. Either pair would have orbited each other for billions of years, but such systems are unstable. As they radiate energy in gravitational waves, they spiral very slowly toward each other, until eventually they collide in the most violent and energetic events known.

Einstein's special theory of relativity tells us that all light travels at the same speed, no matter what its frequency. The gamma-ray bursts provide a laboratory to test this claim, because they give a very short burst of photons in a wide range of energies. Most important, they can take billions of years to reach us, and herein lies the heart of the experiment.

Suppose Einstein is wrong, and photons of different energies travel at slightly different speeds. If two photons created in the same distant explosion were to arrive on Earth at different times, this would surely indicate a breakdown in the special theory of relativity.

What would such a momentous discovery imply? This would first depend on the physical scale at which the breakdown occurred. One place where we do expect special relativity to crumble is at the Planck length. Recall from the preceding chapter that the Planck scale is about 10^{-20} times the size of a proton. Quantum theory tells us that this scale represents a threshold below which the classical picture of spacetime disintegrates. Einstein's special theory of relativity is part of that classical picture, so we might expect it to break down at just that point.

Could any experiment see the effect of a breakdown in the structure of space and time at the Planck scale? Given modern electronics, very tiny differences in the arrival times of photons can be detected, but are modern electronics good enough to measure the even more minuscule effects of quantum gravity? For decades, we theorists have been teaching that the Planck scale is so small that no currently feasible experiment could detect it. Just as most physics professors a hundred years ago held that atoms were too small to see, we have told this lie in countless papers and lectures. And it is a lie.

Remarkably, it took until the mid 1990s for us to realize that we could indeed probe the Planck scale. As sometimes happens, a few people recognized it but were in effect shouted down when they tried to publish their ideas. One was the Spanish physicist Luis Gonzalez-Mestres, of the Centre National de la Recherche Scientifique, in Paris. A discovery like this may be made several times independently until someone brings the point to the attention of the community of specialists in a way that sticks. In this case, it was Giovanni Amelino-Camelia, of the University of Rome. Now in his early forties, Amelino-Camelia is driven, focused, and passionate about physics, with all the charm and fire one associates with a southern Italian. The quantum-gravity community is lucky to count him as a member.

When Amelino-Camelia was a postdoc at Oxford, he set himself the task of looking for a way to observe physics at the Planck scale. This seemed a completely crazy ambition at the time, but he challenged himself to prove common knowledge wrong and come up with some way to do it. He was inspired by the tests of proton decay. Proton decay (see chapter 4) was predicted to be an extremely rare event, but if you got enough protons together, you could expect to see it happen. The huge number of protons would function as an amplifier, making visible something extremely tiny and rare. The question Amelino-Camelia asked himself was whether any such amplifier could enable him to detect phenomena at the Planck scale.

We have already noted two examples of useful amplification: cosmic rays and photons from gamma ray bursts. In both cases, we use the universe itself as an amplifier. Its very size amplifies the probability of extremely rare events, and the enormous amount of time that light takes to travel across it can amplify tiny effects. That these kinds of experiments could theoretically signal a breakdown of special relativity had been pointed out before. What Amelino-Camelia discovered was that we could actually devise experiments to probe the Planck scale, and hence quantum gravity.

A typical change in the speed of a photon due to quantum gravity would be incredibly tiny, but the effect is greatly amplified by the travel time from the gamma-ray burst, which can be billions of years. Physicists realized a few years ago, using rough estimates of the size of quantum gravity effects, that the time between the arrival of photons of different energy that had been traveling this long would be about $1/1,000$ of a second. This is a tiny length of time, but it is well within the range that can be measured with modern electronics. Indeed, the newest gamma-ray detector, called GLAST (for Gamma Ray Large Area Space Telescope), has this kind of sensitivity. It is scheduled to launch in the summer of 2007 and its results are eagerly awaited.

Since the barrier was first broken by Amelino-Camelia and his collaborators, we have discovered that there are many ways to probe the Planck scale with real experiments. Amelino-Camelia's crazy question has become a respectable field of science.

So suppose a new experimental result contradicts special relativity at the Planck scale. What would this tell us about the nature of space and time?

I mentioned at the beginning of this chapter that there were two possibilities. We have already discussed one, which is that the principle of the relativity of motion is wrong — meaning that we could indeed distinguish absolute motion from absolute rest. This would reverse a principle that has been the linchpin of physics since Galileo. I personally find this possibility abhorrent, but as a scientist I must acknowledge that it is a real possibility. Indeed, if the results of AGASA, the Japanese cosmic-ray experiment, hold up, such a breakdown in special relativity may have already been seen.

But is this the only possibility? Most physicists would probably say that if photons with different energies travel at different speeds, then special relativity is wrong. I would certainly have said this a decade ago. But I would have been mistaken.

Einstein's special theory of relativity is based on two postulates: One is the relativity of motion, and the second is the constancy and universality of the speed of light. Could the first postulate be true and the other false? If that was not possible, Einstein would not have had to make two postulates. But I don't think many people realized until recently that you could have a consistent theory in which you changed only the second postulate. It turns out that you can, and working this out has been one of the most exciting things I've had the good fortune to participate in during my career.

The new theory is called *deformed* or *doubly special relativity* — DSR for short. It came from asking a simple question, which seems to lead to a paradox.

As noted, the Planck length is believed to be a kind of threshold below which is revealed a new kind of geometry, one that is intrinsically quantum mechanical. Different approaches to quantum gravity agree on one thing: The Planck length is in some sense the size of the smallest thing that can be observed. The question is, will *all observers* agree on what this shortest length is?

According to Einstein's special theory of relativity, different observers disagree on the lengths of moving objects. An observer riding on the meter stick will say it is a meter long. But any observer moving with respect to it will observe it to be shorter. Einstein called this *the phenomenon of length contraction.*

But this implies that there cannot *be* such a thing as a "shortest length." No matter how short something is, you can make it still shorter by moving relative to it very close to the speed of light. Thus there appears to be a contradiction between the idea of the Planck length and special relativity.

Now, you might think that anyone professionally involved in the problem of quantum gravity would have stumbled on this contradiction. You might even think that some bright undergraduate in a first-year physics course would have raised the question. After all, each of the brilliant physicists responsible for the most difficult work in string theory and quantum gravity was once a naïve student. Wouldn't at least a few have seen the problem? But to my knowledge very few did, until recently.

One who did was Giovanni Amelino Camelia. At some point in 1999, he came upon the paradox just described, and he solved it. His idea was to extend the reasoning that had led Einstein to special relativity.

The second postulate of special relativity, which says that the

speed of light is universal, appears to be almost contradictory in itself. Why? Consider a single photon, tracked by two observers. Assume that the two observers move with respect to each other. If
they measure the speed of that single photon, we would normally
expect them to get different answers, because this is the way normal
objects behave. If I see a bus pull ahead of me at what looks to me
like a speed of 10 kilometers an hour because I am in a car scream-
ing down the highway at 140 kilometers per hour, an observer
standing on the side of the road will see the bus moving at 150
km/hour. But if I observe a photon under the same circumstances,
special relativity says that the roadside observer will measure the
photon to have the same speed that I think it has.

So why is this not a contradiction? The key is that we do not
measure speed directly. Speed is a ratio: It is a certain distance per
a certain time. The central realization of Einstein is that different
observers measure a photon to have the same speed, even if they
are moving with respect to each other, because they measure space
and time differently. Their measurements of time and distance
vary from each other in such a way that one speed, that of light, is
universal.

But if we can do this for one constant, why not for another? Could
we play the trick for distance as well? That is, we understand that,
generally, observers measure a moving meter stick to be less than a
meter long. This will be true for most lengths, but can we arrange
things so that when we finally get all the way down to the Planck
length, the effect goes away? This means that if a stick is exactly a
Planck-length long, all observers will agree on its length, even if it
is moving. Could we then have two universal quantities, a speed
and a length?

Einstein got away with the first trick because nothing can go
faster than light. There are two kinds of things in the world —
things that go the speed of light and things that go less than the
speed of light. If one observer sees something go less than the speed
of light, all observers will. And if one observer sees something go exactly the speed of light, all observers will agree about that, too.

Amelino-Camelia's idea was to play the same game with length.
He proposed modifying the rules by which space and time measurements differ from one observer to another, so that if something is

the Planck length, then all observers will agree it is a Planck-length long, and if it is longer than that, all observers will agree about that, too. This scheme can be consistent, because nothing can be smaller than the Planck length, for any observer.

Amelino-Camelia quickly found that there is a modification of Einstein's special-relativity equations that realizes this idea. He called it *doubly special relativity*, because the trick that made relativity special had now been played twice. I had been following his efforts to invent ways to probe the Planck scale, but in 2000, when he circulated a preprint on the idea of doubly special relativity, I didn't at first understand it.[1]

That's embarrassing enough, but here's something even more embarrassing. About ten years earlier, I had run into the very same paradox. It arose in work I was doing on a theory of quantum gravity called *loop quantum gravity*. The details aren't important — the point is that our calculations in loop quantum gravity appeared to contradict Einstein's special theory of relativity. Now I understand that those particular calculations actually did contradict Einstein's special theory. But at the time, that possibility was too scary to contemplate, and after struggling with this, I dropped the whole line of research. Indeed, this was the first in a series of steps that eventually led me to forsake loop quantum gravity and work for a time on string theory.

But just before I dropped it, I had a thought: Perhaps special relativity could be modified so that all observers, moving or not, agree about what the Planck length is. This was the key idea of doubly special relativity, although I wasn't imaginative enough to do anything about it. I thought about it for a bit, couldn't make any sense of it, and then went on to something else. Even seeing Amelino-Camelia's paper ten years later did not bring it back to me. I had to come to the idea from another direction. At this time, I was a visiting professor at Imperial College in London, and I was getting to know a remarkable scientist there named João Magueijo, a bright young cosmologist from Portugal, about the same age as Giovanni Amelino-Camelia and with an equally intense Latin temperament.

João Magueijo was known for having a really crazy idea, which was that light traveled faster in the very early universe. This idea makes inflation unnecessary, because it explains how every region of the early universe could have been in causal contact and thus at the same temperature. There would have been no need for an exponential expansion in the earliest moments in order to bring this about.

That's nice, but the idea is nuts — really nuts. It disagrees with both special and general relativity. There is no other word for it but "heretical." However, the British academic world has a soft spot for heretics, and Magueijo was thriving at Imperial College. Had he been in the United States, I doubt he would even have been hired as a postdoc with an idea like that.

Magueijo had developed his idea with a young professor at Imperial named Andreas Albrecht, who as a graduate student at the University of Pennsylvania had been one of the inventors of inflation. Albrecht had recently left England to move back to America. After I had been at Imperial for a few months, I found Magueijo at my door. He wanted to see whether there was a way to make his idea of a variable-speed-of-light (VSL) cosmology consistent with special and general relativity. Somehow he felt that talking with me could help.

I didn't know at the time that this had already been done. Indeed, the whole VSL cosmology had been developed earlier, by that imaginative Toronto physics professor John Moffat. A heretic many times over, Moffat had invented the idea and worked it out in a way that was consistent with both special and general relativity, but his attempts to publish his theory in a scientific journal had met with rejection.

As João tells the story in his 2003 book, *Faster Than the Speed of Light,* he learned of Moffat's work when he and Albrecht tried to publish their own paper.[2] It is characteristic of João that his response was to embrace Moffat as a friend — and, indeed, they remain close. He knew of Moffat's work by the time he started to talk with me, but I don't think he understood that it had solved the problem he was trying to solve. Or if he did, he didn't like the way this had been done.

John Moffat is now a friend and colleague of mine at the Perimeter Institute for Theoretical Physics. There is no one I respect more for his boldness and originality. I've also said how much I admire

Giovanni Amelino-Camelia for his revelations about probing the Planck scale. So it pains me to admit that João and I ignored the work of both of them. In a sense, it's good we did, for we found a different solution to the problem of how to make a variable speed of light consistent with the principles of relativity. I certainly would not have tried if I had known that the problem had already been solved — and not once, but twice.

João came to me often with this problem. I always made time to talk with him, because I was attracted by his energy and his fresh way of seeing physics. But for many months, I didn't think very hard about what he was saying. The turning point came when he showed me an old book in which the problem was discussed. It was a textbook on general relativity by a famous Russian mathematical physicist named Vladimir Fock.[3] I knew some of Fock's work in quantum field theory (all physicists do), but I had never seen his book on relativity. The problem João was trying to get me to think about was a homework problem in Fock's book. Once I saw it, I recalled my idea from ten years earlier, and the whole thing came together. The key was indeed to keep the principles of Einstein's special theory but to change the rules so that all observers agree that both the speed of light *and* the Planck scale are universal. Actually, the speed that is constant is no longer the speed of all photons, only very low-energy ones.

At first we didn't see what to do with this idea. We had a story, with some pieces of the math, but not yet a full a theory. At about this time, I went on a trip that included a stop in Rome, where I spent many hours talking with Giovanni Amelino-Camelia. All of a sudden I understood what he was saying. He had come to the same idea we were developing, and he had come to it earlier and worked it out first. Nevertheless, there was a lot about the way he had worked out the idea that I didn't understand. The math seemed complicated, and it appeared to be tied to a formalism invented some ten years earlier by a group of Polish mathematical physicists — a formalism that I certainly couldn't penetrate.

It would take me many years to appreciate the mathematical subtleties of the subject. I found it impenetrable until I started reading early papers by an English mathematician named Shahn Majid, who was one of the inventors of quantum groups. His work was closely related to the mathematics the Polish group was using. Majid had begun with some visionary ideas about how to express the essential insights of relativity and quantum theory in a single mathematical structure. This had led him to quantum groups (which are a revolutionary extension of the idea of a symmetry) and then to modifications of relativity theory based on a subject we call noncommutative geometry. His insights are at the core of the mathematics required to express DSR clearly, but they were lost — at least, to me — in the complicated papers where I had first seen them expressed.

In any case, João and I ignored mathematics and kept talking about physics. Our progress was interrupted by my move to Canada, to the newly founded Perimeter Institute, in September of 2001. A month later, João came to Perimeter as its second visitor. The theory finally fell into place the afternoon after he arrived. We were working in a café in uptown Waterloo called the Symposium, with comfortable couches. He was jetlagged. I was traumatized and exhausted, having just returned from a weekend in New York following the events of September 11. I fell asleep as João was talking, then woke up to find him dozing. I remembered something he had said as I was losing consciousness, and I played with it on a pad, then fell asleep again. I woke up when he started talking, and we had a few mutually lucid minutes before he fell asleep again. And so the afternoon went, as we talked, calculated, and dozed in turn. I can only imagine what the café staff thought. But at some point during that afternoon, we hit on a key factor that had evaded us for months, having to do with trading momenta for positions. When we were done, we had invented a second version of DSR, much simpler than the one developed by Giovanni Amelino-Camelia. Now it is known to experts as DSR II.

This was roughly what João had wanted. In our version, photons that have more energy travel faster. Thus, in the very early universe, when the temperature was very high, the speed of light was, on average, faster than it is now. As you go farther back in time and the temperature approaches the Planck energy, the speed of light becomes infinite. It took somewhat longer to show that this led to a version of a variable-speed-of-light theory that was also consistent with the principles of *general* relativity, but we eventually got there, too. We call this theory *Gravity's Rainbow*, after Thomas Pynchon's novel.

"Doubly special relativity" is a stupid name, but it has stuck. The idea is an elegant one, by now much studied and discussed. We don't know whether it describes nature, but we know enough about it to know that it could.

The first responses to DSR were not encouraging. Some people said it was inconsistent; others said it was nothing more than a very complicated way of writing Einstein's special relativity theory. A few people made both criticisms.

We answered the second criticism by showing that the theory makes predictions different from those made by special relativity. A key role in these discussions was played by a highly cultured fan of heavy-metal music named Jerzy Kowalski-Glikman, from Warsaw. (Perhaps only a European could truly be both.) I believe he was the first person to really comprehend what Giovanni Amelino-Camelia was saying; I certainly understood his papers, which were short and crystal clear, before I understood Giovanni's, which were long, printed in a small font, and full of asides and details. Jerzy found several of the important consequences of doubly special relativity, and it was he who straightened out the relationship between our efforts and the earlier mathematical work of his Polish colleagues.

A watershed in my understanding of DSR and how the various approaches to it were connected was a discussion we had one afternoon at my girlfriend's house in Toronto. Giovanni, Jerzy, João, and I had squeezed ourselves around a small table in her narrow dining room in an attempt to get to the bottom of our disagreements and misunderstandings. Jerzy insisted quietly that if anything was to make sense, it had to fit into a consistent mathematical structure, which for him meant the noncommutative geometries that he and his Polish colleagues had studied. João said that everything to do with physics could be understood without the fancy mathematics. Giovanni argued that it was easy to talk nonsense about these theories if you weren't careful about which mathematical expressions corresponded to things that could be measured. At some point — I don't recall the particular comment that set it off — Giovanni seized the formidable bread knife and howled, "If what you are saying is right, I slit my throat. *Now!*"

We stared at him, and after a moment's shocked silence, we collapsed laughing, and so did he. Only then were we each ready to start listening to what the others were saying.

In fact, there are different versions of DSR, which give different predictions. In some, there is an energy that cannot be exceeded, analogous to the maximal speed of light. In others, there is no maximal energy but there is a maximal momentum. This is unfortunate, as it lessens the predictive power of the theory, but it doesn't seem to take away from the theory's consistency, so it's something we have to live with.

The consistency of DSR was shown by demonstrating that there is a possible universe in which it would be true. The possible universe is like our own, with one difference, which is that space has only two dimensions. It was discovered in the 1980s that quantum gravity can be precisely defined in a world with only two spatial dimensions. We call this *2 + 1 quantum gravity*, for two dimensions of space and one of time. Moreover, if there is no matter, the theory can be exactly solved — that is, one can find exact mathematical expressions that answer any question that can be asked about the world the theory describes.

It turns out that DSR is true in any world with two dimensions of space, quantum gravity, and matter. The particular form of DSR that is realized is the form originally discovered by Giovanni. When Jerzy and I looked back at the literature, we saw that several people had found features of this two-dimensional world that were aspects of DSR, but they had done so before the concept of DSR had been invented. Excited, we described this to Laurent Freidel, a colleague at Perimeter from France who works on quantum gravity. He told us that he not only already knew that but had tried to tell us about it earlier. I'm sure that's true. In discussion, Friedel has more energy than I do, and I usually fail to understand what he is saying, to which he responds by talking faster and louder. In any case, we wrote a paper together that explained why DSR must be true of universes with two dimensions of space.[4]

Sometime after that, Friedel, collaborating with Etera Livine, a French Tahitian postdoc at Perimeter, showed in detail how DSR works out in the theory of 2 + 1 dimensional gravity with matter.[5] These are important results, because the fact that there is a model of a possible world where DSR is true guarantees the consistency of the theory.

There was one more problem that had to be solved before DSR could be considered a viable theory. As noted, in many versions there is a maximal energy that a particle can have, which is usually taken to be the Planck energy. This is no problem experimentally, because the most energy that has been observed is that of the protons in the AGASA cosmic-ray detector, which is about a billionth of this maximum.

But at first glance, it seems that the bound should apply to any sort of body: Not just electrons or protons, but dogs, stars, and soccer balls should all have an energy less than the maximum. This clearly contradicts nature, because any system with more than 10^{19} protons has more energy than the Planck mass. Dogs have about 10^{25} protons, stars even more. We call this the *soccer-ball problem*.

The soccer-ball problem exists in a two-dimensional world, but there is no need to solve it there, since we don't do experiments in that world. It is simply true, in that world, that any object has an energy of less than the Planck energy, no matter how many particles it's composed of.

There is a natural solution to the soccer-ball problem that might hold in our world of three spatial dimensions. João and I proposed this solution early on. The idea is that a body has a maximum energy that is one Planck energy for each proton it contains. Thus, a soccer ball, with about 10^{25} protons, cannot have an energy of more than 10^{25} Planck energies. There is, then, no problem with observation.

We could see that this solution would work, but we did not know why it had to be true. An explanation was recently given by Etera Livine and Florian Girelli, another Perimeter postdoc from France. They found a marvelous way to reformulate the theory so that this solution pops out.[6] Now that the soccer-ball problem has been solved, there is no obstacle I know of to DSR being true of our world. It may well be confirmed by Auger and GLAST observations made over the next few years; if not, it will at least be shown to be false, which means that DSR is a real scientific theory.

We can now return to the question of what the implications would be for different theories of quantum gravity if special relativity broke down. We have seen that such a breakdown can mean two different things, depending on what the experiments tell us. Special relativity could break down completely at this scale, which would mean there really is an absolute distinction between motion and rest. Or special relativity could be preserved but deepened, as in DSR.

Would string theory survive either change? Certainly all known string theories would be proved false, since they depend so heavily on special relativity holding. But might there still be a version of string theory that could be consistent with either type of breakdown? Several string theorists have insisted to me that even if special relativity were seen to break down or be modified, there might someday be invented a form of string theory that could accommodate whatever the experiments see. They're possibly right. String theory has, as noted, many fields that are not observed. There are a lot of ways to change the background of string theory so that there is a preferred state of rest, so that the relativity of motion is wrong. Perhaps in this way a version of string theory could be engineered that agreed with experiment.

What about DSR? Could there be a version of string theory consistent with it? As of this writing, João Magueijo and I are the only people who have investigated this question, and the evidence we found was mixed. We were able to construct a string theory that satisfied some tests for consistency, but we didn't succeed in finding a clear answer with respect to other tests.

So, while all the known versions of string theory are consistent with special relativity, it is also the case that if special relativity proves flawed, string theorists might be able to accommodate such a discovery. What puzzles me is why string theorists think this helps their cause. To me, it's more an indication that string theory is unable to make any predictions because it is no more than a collection of theories, one for each of a vast number of possible backgrounds. The question at issue in the GLAST and Auger observations is the symmetry of space and time. In a background-dependent theory, this is decided by the choice of background. As long as a theory allows it, you can get any answer you need to get by choosing an appropriate background. This is very different from making a prediction.

What about other approaches to quantum gravity? Have any predicted a breakdown of special relativity? In a background-independent theory, the situation is very different, because the geometry of spacetime is not specified by choosing the background. That geometry must emerge as a consequence of solving the theory. A background-independent approach to quantum gravity must make a genuine prediction about the symmetry of space and time.

As I discussed earlier, if the world had two dimensions of space, we know the answer. There is no freedom; the calculations show that particles behave according to DSR. Might the same be true in the real world, with three dimensions of space? My intuition is that it would, and we have results in loop quantum gravity that provide evidence, but not yet proof, for this idea. My fondest hope is that this question can be settled quickly, before the observations tell us what is true. It would be wonderful to get a real prediction out of a quantum theory of gravity and then have it shown to be false by an unambiguous observation. The only thing better would be if experiment confirmed the prediction. Either way, we would be doing real science.

15

Physics After String Theory

In the last two chapters, we saw that there is reason to expect dramatic progress in the search for the laws of nature. There are hints that surprising experimental discoveries may be around the corner. And a far-reaching extension of relativity theory offers predictions for experiments in progress. Whether doubly special relativity is right or not, it is real science, because experiments now under way will either confirm or refute its main predictions.

The theorists and experimentalists whose work I described in the last two chapters have already inaugurated the post-string era in fundamental physics. In this chapter, I will take you on a tour of this new world, highlighting the most promising ideas and developments. Looking beyond string theory, we find a healthy resurgence of fundamental theory done the old-fashioned way — through hard, concentrated thought about basic questions, mindful of developments in both mathematics and experimental physics. In all the frontier fields — quantum gravity, foundations of quantum physics, elementary-particle physics, and cosmology — bold new ideas are evolving in tandem with fascinating new experiments. These initiatives have to be nourished or they'll die on the vine, but they show great promise.

Let us start with a field in which we are seeing rapid progress: approaches to quantum gravity that embrace rather than evade Einstein's great discovery that the geometry of spacetime is dynamical and contingent.

As I have emphasized several times, it is not enough to have a theory with gravitons made from strings wiggling in space. We need a theory about *what makes up space*, a background-independent theory. As described earlier, the success of general relativity demonstrates that the geometry of space is not fixed. It is dynamical and it evolves in time. This is a basic discovery that cannot be reversed, so any further theory must incorporate it. String theory does not, so if string theory is valid, there must lie behind it a more fundamental theory — one that is background-independent. In other words, whether string theory is valid or not, we still have to discover a background-independent theory of quantum gravity.

Luckily, thanks to work over the last twenty years, we know a lot about how to make such a theory. The field of background-independent approaches to quantum gravity took off in 1986, just two years after the first string theory revolution. The catalyst was the publication by the theoretical physicist Abhay Ashtekar, then at Syracuse University, of a reformulation of general relativity that made the equations much simpler.[1] Interestingly enough, it did so by expressing Einstein's theory in a form very close to that of the gauge theories — the theories that underlie the standard model of particle physics.

Unfortunately, most string theorists have paid no attention to the remarkable progress made in the field of quantum gravity these last twenty years, so the two fields have developed separately. This lack of communication may seem strange to an outsider. It certainly seems strange to me, which is why I have done my best to reverse it by arguing to each community the merits of the other. But I can't say I've had much success. The failure to get people who work on the same problem from different perspectives to communicate with each other is part of what led me to believe that physics was in crisis — and to think hard about how to rescue it.

The whole atmosphere of the field of quantum gravity is different from that of string theory. There are no grand theories, no fads or fashions. There are just a few very good people working hard on several closely related ideas. There are several directions being explored, but there are also some unifying ideas that give the field an overall coherence.

The main unifying idea is simple to state: *Don't start with space, or anything moving in space.* Start with something that is purely quantum-mechanical and has, instead of space, some kind of purely quantum structure. If the theory is right, then space must emerge, representing some average properties of the structure — in the same sense that temperature emerges as a representation of the average motion of atoms.

Thus many quantum-gravity theorists believe there is a deeper level of reality, where space does not exist (this is taking background independence to its logical extreme). Since string theory requires the existence of a background-independent theory to make sense, many string theorists have indicated that they agree. In a certain limited sense, if the strong form of the Maldacena conjecture (see chapter 9) turns out to be true, a nine-dimensional geometry will emerge out of a fixed three-dimensional geometry. It is thus not surprising to hear Edward Witten say, as he did in a recent talk at the Kavli Institute for Theoretical Physics at UC Santa Barbara, that "most string theorists suspect that spacetime is an 'emergent phenomenon,' in the language of condensed matter physics."[2]

Some string theorists have finally begun to appreciate this point, and one can only hope they will follow up by studying the concrete results that have already been obtained. But in fact, most people in quantum gravity have in mind something more radical than the Maldacena conjecture.

The starting point is nothing like geometry. What many of us in quantum gravity mean when we say that space is emergent is that the continuum of space is an illusion. Just as the apparent smoothness of water or silk hides the fact that matter is made of discrete atoms, we suspect that the smoothness of space is not real and that space emerges as an approximation of something consisting of building blocks that we can count. In some approaches, it is just assumed that space is made of discrete "atoms"; in others, this assumption is rigorously derived by combining the principles of general relativity and quantum theory.

Another unifying idea is the importance of *causality*. In classical general relativity, the spacetime geometry tells light rays how to propagate. Since nothing can travel faster than light, once you know how light propagates, you can determine which events a particular event might have caused. Given two things that happen, the first can be the cause of the second only if a particle has propagated from the first to the second going at or less than the speed of light. Thus, the spacetime geometry contains information about which events cause which other events. This is referred to as the *causal structure of a spacetime.*

It is not only the case that the spacetime geometry determines what the causal relations are. This can be turned around: Causal relations can determine the spacetime geometry, because most of the information you need to define the geometry of spacetime is fixed, if you know how light travels.

It's easy to talk about space or spacetime emerging from something more fundamental, but those who have tried to develop the idea have found it difficult to realize in practice. Indeed, several early approaches failed. We now believe they failed because they ignored the role that causality plays in spacetime. These days, many of us working on quantum gravity believe that *causality itself is fundamental* — and is thus meaningful even at a level where the notion of space has disappeared.[3]

The most successful approaches to quantum gravity to date combine these three basic ideas: that space is *emergent*, that the more fundamental description is *discrete*, and that this description involves *causality* in a fundamental way.

The current study of quantum gravity is in some respects analogous to physics one hundred years ago, when people believed in atoms but didn't know the details of atomic structure. But despite this ignorance, Ludwig Boltzmann, Einstein, and others could understand quite a lot about matter using only the fact that it was made of atoms. Knowing nothing more than the approximate size of an atom, they were even able to make predictions of observable effects. Similarly, we have been able to derive important results from simple models based only on the three principles of emergence, discreteness, and causality. Given our ignorance about the details, these models make the simplest possible assumptions about the discrete units of spacetime and then see what can come of them. The most successful of these models was invented by Renate Loll and Jan Ambjørn and is called *causal dynamical triangulations*.[4] This is perhaps too technical a name for an approach with a very simple strategy, which is to represent the basic causal processes by simple building blocks, which indeed look like the blocks that children play with (see Fig. 14). It might be called the Buckminster Fuller approach. The basic idea is that a spacetime geometry is made by piling up a large number of blocks, each of which represents a simple causal process. There are a few simple rules that govern how the blocks can be piled up and a simple formula that gives the quantum-mechanical probability for each such model of a quantum spacetime.

One of the rules that Loll and Ambjørn impose is that each quantum spacetime has to be seen as a sequence of possible spaces that succeed one another, like the ticks of a universal clock. The time coordinate, it is argued, is arbitrary, as in general relativity, but the fact that the history of the world can be seen as a succession of geometries that succeed one another in time is not.

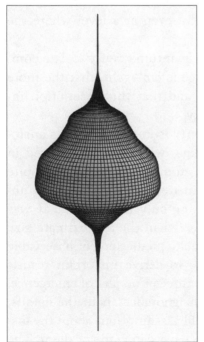

Fig. 14. A model quantum universe according to the causal dynamical triangulation program. The figure depicts the history of a model quantum universe with three dimensions of space, one of which runs horizontally, and one of time, which runs vertically.
Courtesy of Renate Loll

Given that restriction, plus a few simple rules, they have significant evidence that classical spacetime, with its three dimensions of space and one of time, emerges from a simple game of block piling. This is the best evidence yet in a background-independent quantum theory of gravity that a classical spacetime with three dimensions of space can emerge from a purely quantum world based only on discreteness and causality. In particular, it was shown by Ambjørn and others that if no restriction respecting causality is put in, no classical spacetime geometry emerges.

One consequence of these results is that some of the most widely believed ideas about quantum gravity are in fact wrong. For example, Stephen Hawking and others used to argue that causal structure was inessential, and that calculations could be done in quantum gravity by ignoring the differences between time and space — differences that exist even in relativity theory — and treating time as if it were just another dimension of space. This was what Hawking meant by those mysterious references, in his *Brief History of Time*, to time being "imaginary." Ambjørn and Loll's results show that this idea is wrong.

Before their work, other people had investigated the idea that the fundamental building blocks of spacetime involved causality, but no one had hit upon a theory from which classical spacetime could be shown to emerge. One such formulation, called *causal set theory*, took the fundamental units of spacetime to be naked events, whose only attributes were lists of other events that could have caused them and that they could have caused. These ideas are even simpler than Loll and Ambjørn's models, because there is no requirement for a global succession in time. It has so far not been possible to show the emergence of classical spacetime from this theory.

There has been one major triumph of causal set theory, however, which is that it seems to solve the cosmological constant problem. By simply assuming that a classical world emerges from causal set theory, the Syracuse University physicist Rafael D. Sorkin and collaborators predicted that the cosmological constant would be about as small as has since been observed. As far as I'm aware, this is so far the only clean solution to the cosmological constant problem. This solution alone, plus the attraction of a theory based on such simple assumptions, makes this a research program that deserves continued support.

The English mathematical physicist Roger Penrose has also proposed an approach to quantum spacetime based on the principle that what is really fundamental is relations of causality. His approach is called *twistor theory*. He and a few adherents have been working on this since the 1960s. It is based on a reversal of the usual way of seeing events in spacetime. Traditionally, one sees what happens as primary and the relationships between what happens as secondary. Thus the events are real and the causal relations between the events are simply properties of the events. Penrose found that this way of looking at things can be reversed. You can take the elementary causal processes as fundamental and then define events in terms of coincidences between causal processes. More specifically, you can make a new space, consisting of all the light rays in spacetime. You can then translate all of physics into this space of light rays. The result is an incredibly beautiful construction, which Penrose calls *twistor space.*

For the first twenty years after Penrose proposed it, twistor theory developed rapidly. In surprising and beautiful ways, many of the basic equations of physics could be rewritten in terms of twistor space. It really did seem as if you could see the light rays as the most fundamental thing, with space and time just an aspect of relations among them. There was also progress in unification, because equations describing the various kinds of particles take on the same simple form when written in terms of twistor space. Twistor theory partly realizes the idea that spacetime may emerge from another structure. The events of our spacetime turn out to be certain surfaces suspended in the twistor space. The geometry of our spacetime also emerges from structures in twistor space.

But there are problems with this picture. The main one is that twistor space is understood only in the absence of quantum theory. And while twistor space is very different from spacetime, it is a smooth geometrical structure. No one yet knows what a quantum twistor space looks like. Whether quantum twistor theory will make sense, and whether spacetime will emerge from it, has yet to be shown.

The center of twistor theory in the 1970s was Oxford, and I was one of many who were drawn to spend time there. I found a heady atmosphere, not unlike the atmosphere that would develop later at centers of string theory. Penrose was deeply admired, as Edward Witten would be later. I encountered extremely talented young physicists and mathematicians who passionately believed in twistor theory. Several have gone on to prominence as mathematicians.

Twistor theory certainly led to important advances in mathematics. It gave us a deeper understanding of several of the major equations of physics, including the main equations of Yang-Mills theory, which is the basis of the standard model of particle physics. Twistor theory also gave us a deep and stunningly beautiful understanding of a certain set of solutions to Einstein's general theory of relativity. These insights have figured significantly in several different developments, including loop quantum gravity.

But twistor theory has so far not blossomed into a viable approach to quantum gravity — chiefly because it hasn't found a way to incorporate most of general relativity. Still, Penrose and a few colleagues haven't given up. And a few string theorists, led by Witten, have recently begun working on it, bringing to twistor space some new methods that have moved things forward rapidly. This approach does not so far appear to help twistor theory develop into a quantum theory of gravity, but it is revolutionizing the study of gauge theories — evidence, if any were needed, that it was wrong to neglect twistor theory for so long.

Roger Penrose is not the only first-rate mathematician to invent his own approach to quantum gravity. Perhaps the greatest living mathematician — and certainly the funniest — is Alain Connes, who is the son of a chief of detectives from Marseille and has worked for most of his life in Paris. I love to talk to Alain. I don't always understand everything he says, but I go away giddy, both from the profundity of his ideas and the absurdity of his jokes. (These tend to be R-rated, even when they are about black holes or pesky Calabi-Yau manifolds.) Once he broke up a conference talk on quantum cosmology by insisting that to show respect, we should all stand up each time the universe was mentioned. But if I don't always understand Alain, he always understands me; he is one of those people who think so fast that they finish your sentences for you and inevitably improve on what you were about to say. Yet he is so relaxed and confident in himself and his ideas that he is not the least bit competitive, and he is genuinely curious about the ideas of others.

Alain's approach to quantum gravity has been to go back to the foundations and invent a new mathematics that perfectly unifies the mathematical structures of geometry and quantum theory. This is the math I alluded to in chapter 14, called *noncommutative geometry*. "Noncommutative" refers to the fact that quantities in quantum theory are represented by objects that do not commute: That is, AB is not equal to BA. The noncommutivity of quantum theory is closely tied to the fact that you can't measure a particle's position and momentum at once. When two quantities don't commute, you can't know their values simultaneously. Now, this seems counter to the essence of geometry, which starts with a visual image of a surface. The very ability to form a visual image implies complete definition and complete knowledge. To make a version of something like geometry built on things that cannot be simultaneously known was a profound step indeed. What is compelling about it is that it offers a new unification of several areas of mathematics, while putting itself forward as the proper math for the next step in physics.

Noncommutative geometry has turned up in several approaches to quantum gravity, including string theory, DSR, and loop quantum gravity. But none of these capture the depth of Connes's original conception, which he and a few mathematicians, mostly in France, continue to develop.[5] The various versions of it that appear in other programs are based on superficial ideas, such as making the coordinates of space and time into noncommuting quantities. Connes's idea is much deeper; it is a unification at the foundations of algebra and geometry. It could only be the invention of someone who does not just exploit mathematics but thinks strategically and creatively about the structure of mathematical knowledge and its future.

Like the old twistor theorists, the few followers Connes has acquired are committed. For a conference at Penn State University on different approaches to quantum gravity, Alain recommended a famous elder French physicist named Daniel Kastler. The gentleman broke his leg in a bicycle accident a week before the conference, but he clambered out of the hospital and got himself to the Marseille airport, arriving just in time to open the proceedings with the following proclamation: "There is one true Alain, and I am his messenger." String theorists aren't the only ones who have their true believers, but the noncommutative geometers surely have a better sense of humor.

One success of noncommutative geometry is that it leads directly to the standard model of particle physics. As Alain and his colleagues discovered, if you take Maxwell's theory of electromagnetism and write it on the simplest possible noncommutative geometry, out pops the Weinberg-Salam model unifying electromagnetism with the weak nuclear force. In other words, the weak interactions, together with the Higgs fields, show up automatically and correctly.

Recall from chapter 2 that one way to tell whether a particular unification is successful is that there is immediately a sense that the idea agrees with nature. The fact that the correct unification of the weak and electromagnetic forces falls out from the simplest version of Connes's idea is compelling. It is the kind of thing that might have happened with string theory but didn't.

There is another set of approaches focusing on how classical spacetime and particle physics could emerge from an underlying discrete structure. These are models developed by condensed-matter physicists, such as Robert Laughlin, of Stanford; Grigori Volovik, of the Helsinki University of Technology; and Xiao-Gang Wen, of MIT. Recently these approaches have been taken up by young people in quantum gravity, such as Olaf Dreyer. These models are primitive, but they do show that aspects of special relativity, such as the universality of an upper speed limit, can emerge from certain kinds of discrete quantum systems. One provocative assertion of Volovik and Dreyer is that the problem of the cosmological constant is solved — because it was never actually a problem in the first place. They claim that the idea that there was a problem was a mistake, a consequence of taking background-dependent theories too seriously. The mistake, they argue, comes from splitting asunder the basic variables of a theory and treating some of them as frozen background and the others as quantum fields.[6] If they are right about this, it's the most important result to have come out of quantum gravity in many years.

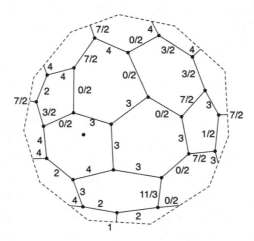

Fig. 15. A spin network, which is a state of quantum geometry in loop quantum gravity and related theories. There are quanta of volume associated with the nodes and quanta of area associated with the edges.

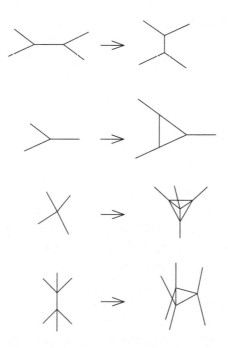

Fig. 16. Spin networks evolve in time through a series of local changes like these.

All the approaches I've been describing are background-independent. Several begin with an assumption that spacetime is composed of discrete building blocks. One would like to do better and show that the discreteness of space and time is a consequence of putting the principles of quantum theory and relativity theory together. This is what *loop quantum gravity* accomplishes. It did so starting with Ashtekar's revolutionary reformulation in 1986 of Einstein's general theory of relativity. What we found was that with no added input, but merely by rewriting Einstein's theory in terms of a new set of variables, it was possible to derive precisely what a quantum spacetime is.

The key idea behind loop quantum gravity is actually an old one, which we have already discussed in chapter 7. It is the idea of a description of a field, like the electromagnetic field, directly in terms of its field lines. (The word "loop" comes from the fact that, in the absence of matter, the field lines can close on themselves, forming a loop.) This was the vision of Holger Nielsen, Alexander Polyakov, and Kenneth Wilson, and it was one of the ideas that led to string theory. Basically, string theory is the development of this visionary idea in a context of a fixed background of space and time. Loop quantum gravity is the same idea but developed in a completely background-independent theory.

This work was made possible by Ashtekar's great discovery that general relativity could be expressed in language like that of a gauge field. The metric of spacetime, then, turns out to be something like an electric field. When we tried to treat the corresponding field lines quantum-mechanically, we were forced to treat them without a background because there was none — the field lines already described the geometry of space. Once we made them quantum-mechanical, there was no classical geometry left. So we had to reinvent quantum field theory in order to work without a background metric. To make a long story short, it took the input of many people, with a variety of skills from physics and mathematics, but we succeeded. The result is loop quantum gravity.

The resulting picture is very simple. A quantum geometry is a certain kind of graph (see Fig. 15). A quantum spacetime is a sequence of events in which the graph evolves by local changes in its structure. This is best illustrated by examples, which are shown in Fig. 16.

The theory leads to many successes. It has been proved finite, in three senses:

1. Quantum geometry is finite, so that areas and volumes come in discrete units.
2. When you compute the probabilities for the quantum geometries to evolve into different histories, they always come out finite (at least in a certain formulation of the theory called the *Barrett-Crane model*).
3. When the theory is coupled to a matter theory, such as the standard model of particle physics, the infinities that ordinarily occur are rendered finite: That is, without gravity, you have to carry out a special procedure to isolate the infinite expressions and render them unobservable; with gravity, there simply are no infinite expressions.

It should be emphasized that there is no uncertainty associated with the foregoing statements. The main results of loop quantum gravity have been proved by rigorous theorems.

The biggest challenge facing loop quantum gravity has from the beginning been to explain how classical spacetime emerges. In the last few years, there has been major progress on this problem, partly thanks to the invention of new approximation procedures. These showed that the theory has quantum states describing universes where the geometry is, to a good approximation, classical. An important step was taken last year by Carlo Rovelli, of the Centre de Physique Théorique in Marseille, and his colleagues, in which they found strong evidence that loop quantum gravity predicts that two masses will attract each other in precisely the way that Newton's law specifies.[7] These results also indicate that at low energies the theory has gravitons, so loop quantum gravity is indeed a theory of gravity.

A lot of effort is now going into applying loop quantum gravity to real-world phenomena. There is a precise description of black-hole horizons that gets the entropy right. These results agree with Bekenstein's and Hawking's old predictions that black holes have entropy and temperature (see chapter 6). As I write, one hot topic among graduate students and postdocs is to predict modifications of Hawking's result for the thermodynamics of black holes that, were they to be measured in some future study of a physical black hole, could confirm or falsify loop quantum gravity.

Loop quantum gravity has also been the basis for models that allow the strongly time-varying geometries inside black holes to be studied. Several calculations give evidence that the singularities inside black holes are removed. Thus time can continue beyond the point at which classical general relativity predicted it must end. Where does it go? It seems to go into newly created regions of space-time. The singularity is replaced by what we call a *spacetime bounce.* Just before the bounce, the matter inside the black hole was contracting. Just after the bounce, it is expanding, but into a new region that did not exist before. This is a very satisfying result, as it confirms an earlier speculation of Bryce DeWitt and John Archibald Wheeler. The same techniques have been used to study what happens in the very early universe. Again, theorists find evidence that the singularity is eliminated, which would mean that the universe existed before the Big Bang.

The elimination of the singularity in black holes provides a natural answer to the black-hole information paradox of Hawking. As noted in chapter 6, the information is not lost; it goes into the new region of spacetime.

The control that loop quantum gravity has given us over the very early universe has made it possible to compute predictions for real observations. Two postdocs at the Perimeter Institute, Stefan Hofmann and Oliver Winkler, were recently able to derive precise predictions for quantum-gravity effects that may be seen in future observations of the cosmic microwave background.[8]

Theorists are also busy trying to predict what we might see in the Auger and GLAST experiments, both of which will indicate whether special relativity breaks down at Planck energies. One great advantage of background-independent approaches is their ability to make predictions for such experiments. Is the principle of the relativity of inertial frames preserved or broken? Is it modified, as in DSR theories? As I have emphasized, no background-dependent theory can make a real prediction for these experiments, because the question is already answered by the choice of background. String theory, in particular, assumes that the relativity of inertial frames remains true in the form originally given by Einstein in his special theory of relativity. Only a background-independent approach can make a prediction for the fate of the principles of special relativity, because the properties of classical spacetime emerge as the solution to a dynamical problem.

Loop quantum gravity promises to be able to make a sure prediction. In the models in which space has only two dimensions, it has already done so: It predicts that DSR is right. There are indications that the same prediction holds for our three-dimensional world, but so far there is no convincing proof of this.

What about the other big problems, such as unification of the particles and forces? Until recently, we thought that loop quantum gravity had little to say about problems other than quantum gravity. We could put matter in the theory and the good results would not be changed. If we wanted, we could put in the entire standard model of particle physics — or any other model of particle physics we wanted to study — but we didn't think loop quantum gravity had anything specific to contribute to the problem of unification. Lately we have realized that we were wrong about this. Loop quantum gravity already has elementary particles in it, and recent results suggest that this is exactly the right particle physics: the standard model.

Last year, Fotini Markopoulou proposed a new way to approach the problem of how the geometry of space could emerge from a more fundamental theory. Markopoulou is the young physicist working on quantum gravity who most often surprises me with unlikely ideas that turn out to be right, and this was one of her best. Rather than asking directly whether or not the geometry of quantum spacetime can emerge as a classical spacetime, she proposed a different approach, based on identifying and studying the motion of particles in quantum geometry. Her idea was that a particle must be some kind of emergent excitation of quantum geometry, traveling through that geometry much as a wave travels through a solid or a liquid. However, in order for the physics we know to be reproduced, these emergent particles have to be describable as pure quantum particles, ignoring the quantum geometry through which they travel.[9]

Normally when a particle is in interaction with an environment, information about its state dissipates into the environment — we say that it *decoheres*. It's difficult to prevent this decoherence from happening; this, by the way, is why it's hard to make a quantum computer, which depends for its efficacy on a particle's being in a pure quantum state. The people who make quantum computers have ideas about when a quantum system will stay pure even in contact with an environment. While working with experts in this area, Markopoulou realized that their insights applied to the problem of how a quantum particle could emerge from a quantum spacetime. She pointed out that to draw predictions from theories of quantum gravity, you can identify such a quantum particle and show it moving as if it were in ordinary space. In her analogy, the environment is the quantum spacetime, which, being dynamical, is constantly changing. The quantum particle must move through it as though it were a fixed, nondynamical background.

Using these ideas, Markopoulou and her collaborators could show that some background-independent theories of quantum gravity have emergent particles. But what are these particles? Do they correspond to anything that has been observed?

At first the problem seemed difficult, because the quantum geometries predicted by loop quantum gravity are very complicated. The particle states are associated with graphs drawn in three-dimensional space. The space is a background, but it has no properties except for its topology; all the information about measures of geometry — like lengths, areas, and volumes — come from the graphs. But because the graphs have to be drawn in space, the theory has a lot of extra information in it that seems to have nothing to do with geometry. This is because of the infinite number of ways the edges of a graph can knot, link, and braid in three-dimensional space.

What is the significance of the knotting, linking, and braiding of the graphs? This question has been with us since around 1988. All this time, we had no idea what the knotting, linking, and braiding meant. Markopoulou saw that emergent particles are coded in these topological structures.

Last spring, I happened to see a preprint by a young Australian particle physicist called Sundance O. Bilson-Thompson. In it he presented a simple braiding of ribbons that quite remarkably captured precisely the structure of the preon models of particle physics I discussed in chapter 5. (Recall that these are models positing the hypothetical particles called preons as the fundamental constituents of protons, neutrons, and other particles the standard model deems elementary.) In his model, a preon is a ribbon, and the various kinds of preons correspond to the ribbon being twisted to the left, right, or not at all. Three ribbons can be braided together, and the various ways to do this correspond precisely to the various particles of the standard model.[10]

As soon as I read the paper, I knew this was the missing idea, because the braids Bilson-Thompson studied could all occur in loop quantum gravity. This meant that the different ways to braid and knot the edges of the graphs in a quantum spacetime must be different kinds of elementary particles. So loop quantum gravity is not just about quantum spacetime — it already has elementary-particle physics in it. And if we could discover Bilson-Thompson's game working precisely in the theory, it would have the *right* elementary-particle physics. I asked Markopoulou if his braids could be her coherent excitations. We invited Bilson-Thompson to collaborate with us, and after several false starts we saw that the argument indeed worked all the way through. Making some mild assumptions, we found a preon model describing the simplest of these particle-like states in a class of quantum-gravity theories.[11]

This result raises many questions, and answering them is now my primary goal. It is too early to tell if it works well enough to give unambiguous predictions for the upcoming experiments at the Large Hadron Collider at CERN. But one thing is clear. String theory is no longer the only approach to quantum gravity that also unifies the elementary particles. Markopoulou's results suggest that many of the background-independent quantum theories of gravity have elementary particles in them as emergent states. And a given theory does not lead to a vast landscape of possible theories. Rather, it shows promise of leading to unique predictions, which will either be in agreement with experiment or not. Most important, this obviates the need to revise the scientific method by invoking the anthropic principle, as Leonard Susskind and others have advocated (see chapter 11). Science done the old-fashioned way is moving ahead.

Plainly, there *are* different approaches to the five fundamental problems in physics. The field of fundamental physics beyond string theory is progressing rapidly, and in several directions, including but not limited to causal dynamical triangulations and loop quantum gravity. As in any healthy field of science, there is a lively interaction with both experiment and mathematics. While there aren't as many people (perhaps two hundred, all told) in these research programs as there are in string theory, it's still quite a lot of people to be tackling foundational problems on the frontiers of science. The big leaps of the twentieth century were made by far fewer. When it comes to revolutionizing science, what matters is quality of thought, not quantity of true believers.

I want to be clear, though, that there is nothing in this new, post-string atmosphere that excludes the study of string theory per se. The idea it is based on — the duality of fields and strings — is, as I have pointed out, one shared with loop quantum gravity. What has led to the present crisis in physics is not this core idea but a particular kind of realization of it, worked out in a background-dependent context — a context that ties it to risky proposals such as supersymmetry and higher dimensions. There is no reason why a different approach to string theory — one more in tune with foundational issues like background independence and the problems in quantum theory — might not be part of the final story. But to find this out, string theory needs to be developed in an open atmosphere, in which it is considered as one idea among several, without any presuppositions as to its ultimate success or failure. What the new spirit of physics cannot tolerate is a presumption that one idea has to succeed, whatever the evidence.

While there is today an exciting sense of progress among quantum-gravity theorists, there is also a strong expectation that the road ahead will bring at least a few surprises. Unlike string theorists in the exhilarating days of the two superstring revolutions, few of the people working on quantum gravity believe they have their hands on a final theory. We recognize that the accomplishments of background-independent approaches to quantum gravity are a necessary step in finishing Einstein's revolution. They show that there can be a consistent mathematical and conceptual language that unifies quantum theory and general relativity. This gives us something string theory does not, which is a possible framework in which to formulate the theory that solves all five of the problems I listed in chapter 1. But we are also fairly sure that we do not yet have all the pieces. Even with the recent successes, no idea yet has that absolute ring of truth.

When you look back at the history of physics, one thing sticks out: When the right theory is finally proposed, it triumphs quickly. The few really good ideas about unification appear in a form that is compelling, simple, and unique; they do not come with a list of options or adjustable features. Newtonian mechanics is defined by three simple laws, Newtonian gravity by a simple formula with one constant. Special relativity was complete on arrival. It may have taken twenty-five years to fully formulate quantum mechanics, but from the beginning it was developed in concert with experiment. Many of the key papers in the subject from 1900 on either explained a recent experimental result or made a definite prediction for an experiment that was shortly done. The same was true of general relativity.

Thus, all the theories that triumphed had consequences for experiment that were simple to work out and could be tested within a few years. This does not mean that the theories could be solved exactly — most theories never are. But it does mean that physical insight led immediately to a prediction of a new physical effect.

Whatever else one says about string theory, loop quantum gravity, and other approaches, they have not delivered on that front. The standard excuse has been that experiments on this scale are impossible to perform — but, as we've seen, such is not the case. So there must be another reason. I believe there is something basic we are all missing, some wrong assumption we are all making. If this is so, then we need to isolate the wrong assumption and replace it with a new idea.

What could that wrong assumption be? My guess is that it involves two things: the foundations of quantum mechanics and the nature of time. We have already discussed the first; I find it hopeful that new ideas about quantum mechanics have been proposed recently, motivated by studies of quantum gravity. But I strongly suspect that the key is time. More and more, I have the feeling that quantum theory and general relativity are both deeply wrong about the nature of time. It is not enough to combine them. There is a deeper problem, perhaps going back to the origin of physics.

Around the beginning of the seventeenth century, Descartes and

Galileo both made a most wonderful discovery: You could draw a graph, with one axis being space and the other being time. A motion through space then becomes a curve on the graph (see Fig. 17). In this way, time is represented as if it were another dimension of space. Motion is frozen, and a whole history of constant motion and change is presented to us as something static and unchanging. If I had to guess (and guessing is what I do for a living), this is the scene of the crime.

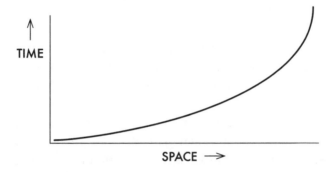

Fig. 17. *Since Descartes and Galileo, a process unfolding in time has been represented as a curve on a graph, with an additional dimension representing time. This "spatialization" of time is useful but may be challenged as representing a static and unchanging world — a frozen, eternal set of mathematical relations.*

We have to find a way to *unfreeze time* — to represent time without turning it into space. I have no idea how to do this. I can't conceive of a mathematics that doesn't represent a world as if it were frozen in eternity. It's terribly hard to represent time, and that's why there's a good chance that this representation is the missing piece.

One thing is clear: I can't get anywhere thinking about this kind of problem within the confines of string theory. Since string theory is limited to the description of strings and branes moving in fixed-background spacetime geometries, it offers nothing for someone who wants to break new ground thinking about the nature of time or of quantum theory. Background-independent approaches offer a better starting point, because they have already transcended the classical picture of space and time. And they are simple to define and easy to play with. There's an added bonus, which is that the mathematics involved is close to the one a few mathematicians have used to explore radical ideas about the nature of time — an area of logic called topos theory.

One thing I do know about the question of how to represent time without its turning into a dimension of space is that it comes up in other fields, from theoretical biology to computer science to law. In an effort to shake free some new ideas, the philosopher Roberto Mangabeira Unger and I recently organized a small workshop at Perimeter bringing together visionaries in each of these fields to talk about time. Those two days were the most exciting I've spent in years.[12]

I won't say more about this, because I want to move on to a different question. Suppose an intellectually ambitious young person with an original and impatient mind wants to think deeply about the five great questions. Given our failure to definitively solve any of them, I can't imagine why such a person would want to be limited to working in any of the current research programs. Clearly, if string theory or loop quantum gravity by themselves were the answer, we would know it by now. They may be starting points, they may be parts of the answer, they may contain necessary lessons. But the right theory must contain new elements, which our ambitious young person is perhaps uniquely qualified to search for.

What has my generation bequeathed to these young scientists? Ideas and techniques they may or may not want to use, together with a cautionary tale of partial success in several directions, resulting in a general failure to finish the job that Einstein started a hundred years ago. The worst thing we could do would be to hold them back by insisting that they work on our ideas. So the question for the last part of the book is a question I ask myself every morning: Are we doing all we can to support and encourage young scientists — and, by virtue of this, ourselves — to transcend what we have done these last thirty years and find the true theory that solves the five great problems of physics?

IV

LEARNING FROM EXPERIENCE

16

How Do You Fight Sociology?

IN THIS LAST PART of the book, I want to return to the questions I raised in the Introduction. Why, despite so much effort by thousands of the most talented and well-trained scientists, has fundamental physics made so little definitive progress in the last twenty-five years? And given that there are promising new directions, what can we do to ensure that the rate of progress is restored to what it was for two centuries before 1980?

One way to describe the trouble with physics is to say that there is no work in theoretical elementary-particle physics over the last three decades that is a sure bet for a Nobel Prize. The reason is that a condition of the prize is that the advance has been checked by experiment. Of course, ideas such as supersymmetry or inflation may be shown by experiment to be true, and if they are, their inventors will deserve Nobel Prizes. But we cannot now say that a discovery of any hypothesis about physics beyond the standard model of particle physics is assured.

The situation was very different when I entered graduate school in 1976. It was abundantly clear that the standard model, which had been put in final form only three years earlier, was a definitive advance. There had already been substantial experimental confirmation of it and more was on the way. There was no serious doubt that its inventors would sooner or later be awarded Nobel Prizes for their work. And, in time, they were.

Nothing like that is true now. There have been many prizes awarded for work in theoretical particle physics in the last twenty-five years, but not the Nobel. The Nobel is not given for being smart or successful; it is given for being right.

This is not to deny that there have been great technical advances in each of the research programs. It is said that there are more scientists working now than in the whole history of science. This is certainly true of physics; there may be more professors of physics in a large university department today than there were a hundred years ago in the whole of Europe, where almost all the advances were being made. All these people are working, and much of the work is technically very sophisticated. Moreover, the technical level of young theoretical physicists is much higher today than it was a generation or two ago. There is more for young people to master, and they somehow manage to do it.

Still, if we judge by the standards of the two hundred years before 1980, it does appear that the pace of irreversible progress in elementary-particle theory has slowed.

We've already discussed the easy explanations for the failure of the last twenty-five years. It's not for lack of data; there are plenty of unexplained results to excite the imaginations of theorists. It's not that theories take a long time to be tested; there's rarely been more than a decade between a new theory's prediction of new phenomena and its confirmation. It's not for lack of effort; far more people now work on problems in fundamental physics than in the whole combined history of the subject. And it certainly cannot be blamed on a lack of talent.

In earlier chapters, I hypothesized that what has failed is not so much a particular theory as a particular style of research. If one spends time in both the community of string theorists and the community of people working on background-independent approaches to quantum gravity, one cannot help but be struck by a great difference in style and in the values expressed by the two communities. These differences reflect a split in theoretical physics that goes back more than half a century.

The style of the quantum-gravity world is inherited from what used to be called the relativity community. This was led by students and associates of Einstein, and by their students in turn — people like Peter Bergmann, Joshua Goldberg, and John Archibald Wheeler. The core values of this community were respect for individual ideas and research programs, suspicion of fashion, a reliance on mathematically clean arguments, and a conviction that the key problems were closely related to foundational issues about the nature of space, time, and the quantum.

The style of the string theory community, on the other hand, is a continuation of the culture of elementary-particle theory. This has always been a more brash, aggressive, and competitive atmosphere, in which theorists vie to respond quickly to new developments (before 1980, these were usually experimental) and are distrustful of philosophical issues. This style supplanted the more reflective, philosophical style that characterized Einstein and the inventors of quantum theory, and it triumphed as the center of science moved to America and the intellectual focus moved from the exploration of fundamental new theories to their application.

Science does need different styles, in order to address different kinds of problems. My hypothesis is that what's wrong with string theory is the fact that it was developed using the elementary-particle-physics style of research, which is ill suited to the discovery of new theoretical frameworks. The style that led to the success of the standard model is also hard to sustain when disconnected from experiment. This competitive, fashion-driven style worked when it was fueled by experimental discoveries but failed when there was nothing driving fashion but the views and tastes of a few prominent individuals.

When I started my studies of physics, in the mid 1970s, both these research styles were healthy. There were many more elementary-particle physicists than relativists, but there was room for both. There were not many places for people who wanted to develop their own solutions to the deep foundational issues about space, time, and the quantum, but there were enough to support the few who had good ideas. Since then, while the need for the relativists' style has grown, their place in the academy has shrunk, due to the dominance of string theory and other large research programs. With the exception of a single research group at Pennsylvania State University, no assistant professors working on an approach to quantum gravity not based on string theory or higher dimensions have been hired by a U.S. research university since around 1990.

Why did the style least suited to the problem at hand come to dominate physics, both here and abroad? This is a sociological question, but it is one we must answer if we are to make constructive suggestions for restoring our discipline to its former vitality.

To put the problem in context, we should look at some pervasive changes in the academic landscape that a young person must negotiate in order to pursue a career in science.

The most striking change is that there is much more pressure on young people to compete for the regard of influential older scientists. The great generation that made American science, now close to retirement, may have had to compete for the top spots in elite universities and institutes, but there was not a lot of pressure if all you wanted was a professorship somewhere that gave you freedom to pursue your work. From the 1940s to the 1970s, the growth of universities was exponential, and it was common for young scientists to have several offers of faculty positions at universities upon graduation. I've met more than one older colleague who never once actually had to apply for a job.

Things are different now. Universities stopped growing in the early 1970s, yet the professors hired in previous eras have continued to train graduate students at a steady rate, which means that there is a significant overproduction of new PhDs in physics and other sciences. As a result, there is fierce competition for places in research universities and colleges at all levels of the academic hierarchy. There is also much more emphasis on hiring faculty who will be funded by the research agencies. This greatly narrows the options for people who want to pursue their own research programs rather than follow those initiated by senior scientists. So there are fewer corners where a creative person can hide, secure in some kind of academic job, and pursue risky and original ideas.

Related to this is the fact that the universities are now much more professionalized than they were a generation or two ago. Whereas university faculties have stopped growing, there has been a marked increase in the number and power of administrators. Thus, in hiring, there is less reliance on the judgment of individual professors and more on statistical measures of achievement, such as funding and citation levels. This also makes it harder for young scientists to buck the mainstream and devote themselves to the invention of new research programs.

In our attempts to make unbiased evaluations of our peers' work, we professors tend almost reflexively to reward those who agree with us and penalize those who disagree. Even when we rise above academic politics, we often fall into the trap of evaluating fellow scientists on the basis of one-dimensional characterizations. In faculty meetings and informal discussions, we talk about who is "good" and who is not, as if we really knew what that meant. Can a person's life work be reduced to "Angela is not as good as Chris"? It often seems as though achievements requiring nothing more than cleverness and hard work are valued more highly than probing thought or imagination. Intellectual fads are far too important, and people who ignore them have dicey academic careers.

I once worked on a project with a retired general who had headed a college for military officers and then become a business consultant. He talked about his frustrations in trying to work with universities. I asked him what he perceived the problem to be. He said, "There is a simple but essential thing we teach to every Marine officer, that no university administrator I've met seems to know: *There is a big difference between management and leadership.* You can manage the procurement of supplies, but you must lead soldiers into battle." I agree with him. In my time in universities, I've seen much more management than leadership.

The problem is of course not confined to science. The pace of innovation in curriculum planning and teaching methods is positively medieval. Any proposal for change has to be approved by the faculty, and in general most professors see nothing wrong with how they have been teaching for decades. I learned early how resistant universities are to change. I was fortunate to have attended a college where the first-year physics course was quantum physics. This is rare. Despite the fact that quantum physics superseded Newtonian mechanics eighty years ago, most colleges and universities in North America still postpone quantum mechanics until the third year of study, and even then it is offered only to physical science majors. Since I knew how to teach a course in freshman quantum mechanics, I proposed doing so as a graduate student at Harvard. I got a young faculty member, Howard Georgi, to agree to teach it with me, but the course was vetoed by the dean of arts and sciences. This had nothing to do with our proposal, he told me, but with the fact that it hadn't gone through the requisite committees. "If we let every professor teach what they wanted to," he said, "we would have educational chaos." I'm not sure that educational chaos is such a bad idea; in any event, Harvard still doesn't have a quantum mechanics course for freshmen.

It is an unfortunate fact that the number of American students graduating with degrees in physics has been declining for decades. You might think this would lessen the competition for physics positions. It doesn't, because the decline in undergraduate degrees is more than compensated for by increases in PhDs earned by bright, ambitious students from the developing world. The same situation obtains in other advanced countries.

I've sometimes had occasion — as a member of a Yale faculty committee formed to investigate the phenomenon, and since then out of my own concern — to ask undergraduates who leave physics why they did so. One reason they give is that the physics curriculum is boring — the first year just repeats what they have learned in high school and there's no sign of exciting topics like quantum theory, cosmology, black holes, and so on. In hopes of helping to reverse the decline in physics students, I have proposed making quantum mechanics the freshman physics course at every university that has employed me. I have been refused each time, although twice I was allowed to teach small demonstrations in quantum theory. These were successful, and a few of the students who took them are now launched on good careers.

My purpose here is not to argue for curriculum reform, but this example suggests that universities don't function well as vehicles for innovation, even when nothing more is at stake than modernizing a curriculum that is eight decades behind the science.

Scientists across the board lament the pace of progress in their field. I know several biologists and experimental physicists who complain bitterly about opportunities squandered because the senior scientists in control of their departments have lost the boldness and imagination they undoubtedly possessed as new PhDs. Good ideas are not taken seriously enough when they come from people of low status in the academic world; conversely, the ideas of high-status people are often taken too seriously.

There is no way we can address these dysfunctional trends without investigating the sociology that has fostered them. If we physicists have the hubris to try to explain the fundamental laws of nature, certainly we ought to be able to think rationally about the sociology of the academy and the counterproductive decision making that plagues our academic institutions.

It is worth noting that the word "sociology" comes up more nowadays among string theorists than among any other group of scientists I know. It seems to be shorthand for "the view of the community." In discussing the current state of affairs with young string theorists, you often hear them say things like "I believe in the theory, but I hate the sociology." If you comment on the narrowness of viewpoints represented at string theory conferences or on the rapid succession of fashionable research topics from one year to the next, a string theorist will agree and add, "I don't like it, but it's just the sociology." More than one friend has advised me that, "The community has decided string theory is right and there is nothing you can do about it. You can't fight sociology."

A real sociologist will tell you that to understand the workings of a community you have to investigate power. Who has power over whom, and how is that power exercised? The sociology of science is not a mysterious force; it refers to the influence that older, established scientists have over the careers of younger scientists. We scientists feel uncomfortable talking about it, because it forces us to confront the possibility that the organization of science may not be entirely objective and rational.

But after thinking about this for a long time, I have become convinced we have to talk about the sociology of theoretical physics, because the phenomena we refer to collectively as its "sociology" are having a significant negative effect on its progress. Even though most string theorists are people of integrity who pursue their work with the best of intentions, there are aspects of the field's sociology that are aberrant, compared with the ideals that define the larger scientific community. These have led to pathologies in the methodology of theoretical physics that delay progress. The issue is not whether string theory is worth doing or should be supported, but why string theory, in spite of a dearth of experimental predictions, has monopolized the resources available to advance fundamental physics, thus choking off the investigation of equally promising alternative approaches. There is good evidence that the progress of string theory itself has been slowed by a sociology that restricts the set of questions investigated and excludes the kind of imaginative and independent-minded scientists that progress requires.

I should point out that there has always been a dominant field within theoretical physics. At one time it was nuclear physics, then it was elementary-particle physics. String theory is only the most recent example. Perhaps the physics community is organized in such a way that there will always be a dominant field at any given moment. If so, then we need to examine why.

The first thing that outsiders notice about the string theory community is its tremendous self-confidence. As a witness to the first superstring revolution, in 1984, I remember the sense of triumph that greeted the new theory. "It will all be over in the next twelve to eighteen months," Dan Friedan, one of the young stars of the field, advised me. "You'd better get in while there's still something left to do in theoretical physics." This was just one of many assertions that things would be wrapped up quickly.

Of course, they weren't. But through all the subsequent highs and lows, many string theorists have continued to be supremely confident both of the truth of string theory and of their superiority over those unable or unwilling to do it. To many string theorists, especially the young ones with no memory of physics before their time, it is incomprehensible that a talented physicist, given the chance, would choose to be anything other than a string theorist.

This attitude of course puts off physicists in other fields. Here are the thoughts of JoAnne Hewett, a particle physicist at the Stanford Linear Accelerator Center, on her blog:

> I find the arrogance of some string theorists astounding, even by physicists' standards. Some truly believe that all non-stringy theorists are inferior scientists. It's all over their letters of recommendation for each other, and I've actually had some of them tell me this to my face. . . . String theory [is perceived to be] so important that it must be practiced at the expense of all other theory. There are two

manifestations of this: String theorists have been hired into faculty positions at a disproportionally high level not necessarily commensurate with ability in all cases, and the younger string theorists are usually not well educated in particle physics. Some literally have a hard time naming the fundamental particles of nature. Both of these manifestations are worrying for the long-term future of our field.[1]

The arrogance Dr. Hewett describes has been a feature of the community of string theorists from the very beginning. Subrahmanyan Chandrasekhar, perhaps the greatest astrophysicist of the twentieth century, loved to tell the story of a visit to Princeton in the mid 1980s, where he was feted in honor of his recent Nobel Prize. At the dinner, he found himself seated next to an earnest young man. As physicists often do to make conversation, he asked his dinner companion, "What are you working on these days?" The reply was, "I work on string theory, which is the most important advance in physics in the twentieth century." The young string theorist went on to advise Chandra to drop what he was doing and switch to string theory or risk becoming as obsolete as those in the 1920s who did not immediately take up quantum theory.

"Young man," Chandra replied, "I knew Werner Heisenberg. I can promise you that Heisenberg would never have been so rude as to tell someone to stop what they were doing and work on quantum theory. And he certainly would never have been so disrespectful as to tell someone who got his PhD fifty years ago that he was about to become obsolete."

Anyone who hangs out with string theorists encounters this kind of supreme confidence regularly. No matter what the problem under discussion, the one option that never comes up (unless introduced by an outsider) is that the theory might simply be wrong. If the discussion veers to the fact that string theory predicts a landscape and hence makes no predictions, some string theorists will rhapsodize about changing the definition of science.

Some string theorists prefer to believe that string theory is too arcane to be understood by human beings, rather than consider the possibility that it might just be wrong. One recent posting on a physics blog laid this out beautifully: "We can't expect a dog to understand quantum mechanics, and it may be that we are reaching the limit of what humans can understand about string theory.

Maybe there are advanced civilizations out there to whom we appear as smart as dogs do to us, and maybe they have figured out string theory well enough to have moved on to a better theory. . . ."[2] Indeed, string theorists seem to have no problem believing that string theory must be right while acknowledging that they have no idea what it really is. In other words, string theory will subsume whatever comes after it. The first time I heard this view expressed, I thought it was a joke, but the fourth iteration convinced me that the speaker was serious. Even Nathan Seiberg, who is a celebrated theorist at the Institute for Advanced Study, was quoted in a recent interview as saying ("with a smile"), "If there is something [beyond string theory], we will call it string theory."[3]

A related characteristic is a sense of entitlement and a lack of regard for those who work on alternative approaches to the problems string theory claims to solve. Indeed, string theorists are normally uninterested in, and often ignorant of, anything that is not labeled string theory. In contrast to the practice at quantum-gravity meetings, the major string theory conferences never invite scientists working on rival approaches to give papers. This of course serves only to bolster the assertions of string theorists that string theory is the only approach yielding successful results on quantum gravity. The disregard of alternative approaches sometimes borders on disdain. At a recent string theory conference, an editor from Cambridge University Press confided to me that a string theorist had told him he would never consider publishing with the press because it had put out a book on loop quantum gravity. This kind of thing is not as rare as it should be.

String theorists are aware of their dominant position in the physics world, and most seem to feel that it's deserved — if the theory itself doesn't justify it, the fact that so many talented people work on it certainly should. If you raise detailed questions about one of string theory's claims with an expert, you risk being regarded, with faint puzzlement, as someone who has inexplicably chosen a path that precludes membership in the club. Of course, this isn't true of the more open-minded string theorists — but there is a peculiar tightening of the face muscles that I've seen too often to ignore, and it usually happens when a young string theorist suddenly realizes that he or she is talking to someone who does not share all the assumptions of the clan.[4]

Another hallmark of string theory is that, unlike other fields of physics, there is a clear distinction between string theorists and non–string theorists. You may have written several papers on string theory, but that does not necessarily mean you will be considered by string theorists to be one of them. At first I found this puzzling. I was pursuing an old strategy of working on different approaches to try to learn from each what I could. I also initially saw much of what I did, even the work on quantum gravity, as addressing an important open issue in string theory, which was how to make it a background-independent formulation. Eventually some friends explained to me that to be considered part of the string theory community — and hence to have any hope of making a mark there — you had to work not just on string theory but on the particular problems that string theorists were then preoccupied with. I don't think it occurred to my friends that doing so might compromise my judgment or impinge on my academic freedom.

I have a broad range of interests, and I've always gone to conferences in fields outside my own. But only at string theory conferences have people come up to me and asked, "What are *you* doing here?" If I explained that I was working on string theory and wanted to see what other people were doing, they would say, brows quizzically furrowed, "But aren't you that loop guy?" No one at a conference on astrophysics, cosmology, biophysics, or postmodernism has ever asked me what I was doing there. At one string theory conference, a leading string theorist sat down, offered his hand, and said, "Welcome home!" Another said, "It's so nice to see you here! We've been worried about you."

In any given year, there are no more than two or three areas in string theory being intensively investigated. These change from year to year, and the fashions can be tracked by looking at the titles of talks at annual major string theory conferences. Often at least two thirds of the talks concern one or two directions that were not strongly represented two years previously and will be almost absent from the conference two years later. Young people are very aware that a successful career requires following two or more of these fads in quick succession, just long enough to get a good postdoc and then a good assistant professorship. If you talk about this with the leaders of string theory, as I have from time to time, you discover that they genuinely believe that concentrating the efforts of a large community of very bright people will lead to faster progress than encouraging colleagues to think independently and pursue a variety of directions.

This monolithic and (as it was called by one senior string theorist) "disciplined" approach has had three unfortunate consequences. First, problems that cannot be solved in two or three years are dropped and often never revisited. The reason is simple and brutal: Young string theorists who do not quickly give up their hard-won specialization in a no-longer-trendy area and switch to the new direction sometimes find themselves without academic positions. Second, the field continues to be driven by the ideas and research agendas of a few people who are now quite senior. In the past decade, only two young string theorists, Juan Maldacena and Raphael Bousso, have made discoveries that changed the direction of the field. This stands in contrast to many other fields in physics, where the majority of new ideas and directions come from people in their twenties and thirties. Third, string theory uses the talents and labor of the large number of people in its community inefficiently. There is much duplication of effort, while many potentially important ideas go unexplored. This narrowing of avenues is evident to anyone who sits on university committees that choose postdocs. In areas such as cosmology, quantum information theory, or quantum gravity, there are as many research proposals as candidates, and often ideas that no one has heard before. In the string theory pool, you tend to encounter the same two or three research proposals over and over again.

Of course, the young people know what they're doing. I've had many years of experience on such committees, and I've found that with a few exceptions, the standards used by string theorists in evaluating their applicants are different from those in other fields. The ability to do mathematically clever work on problems of current interest is, as far as I can tell, valued over the invention of original ideas. Someone who had published papers only with the leading senior scientists and whose research proposal showed little evidence of independent judgment or originality would probably not be offered a position in a leading center of quantum gravity, but it seems the surest route to a postdoc in the leading string theory centers. The kind of applicant that excites me — someone with single-authored papers describing surprising insights and risky new ideas — leaves my string theory friends cold.

In other communities I spend time in, such as quantum gravity and cosmology, there is a diversity of views toward the open problems. If you talk to five different experts, young and old, you will get five different takes on where a subject is headed. Except for recent arguments over the landscape and the anthropic principle, string theorists have maintained a remarkable uniformity of views. One hears the same thing, sometimes in the same words, from different people.

I know some young string theorists who will object to this characterization. They insist that there is a wide range of views within the community — a range that outsiders just aren't privy to. This is good to know, but what people say privately to their friends is not the point. Indeed, if that wider range of views is expressed privately rather than publicly, it suggests that there is a hierarchy controlling the conversation — and the research agenda.

The deliberate narrowing of the research agenda by the leaders of string theory is deplorable not just in principle but also because it has almost certainly led to slower progress. We know this because of the number of ideas that became important for the field many years after they were first proposed. For example, the discovery that string theory is composed of a vast collection of theories was first published by Andrew Strominger in 1986, but it was widely discussed by string theorists only after 2003, following the work of Renata Kallosh and her colleagues at Stanford.[5] Here is a recent quote from Wolfgang Lerche, a well-known string theorist at CERN.

> Well, what I find irritating is that these ideas are out since the mid-80s; in one paper on 4d string constructions a crude estimate of the minimal number of string vacua was made, to the order 10^{1500}; this work had been ignored (because it didn't fit into the philosophy at the time) by the same people who now re-"invent" the landscape, appear in journals in this context and even seem to write books about it. . . . [T]he whole discussion could (and in fact should) have been

[sic] taken place in 1986/87. The main thing what [sic] has changed since then is the mind of certain people, and what you now see is the Stanford propaganda machine working at its fullest.[6]

My own proposal that string theory had to be regarded as a landscape of theories was first published in 1992 and was also ignored.[7] This is not an isolated example. Two eleven-dimensional supersymmetric theories were invented before the first superstring revolution in 1984 but were ignored until they were revived in the second one, more than a decade later. These were eleven-dimensional supergravity and the eleven-dimensional supermembrane. Between 1984 and 1995, a small number of theorists worked on these theories, but they were pushed to the margins of the string theory community. I can recall several derisive references by American string theorists to those "European supergravity fanatics." After 1995, these theories were conjectured to be unified with string theory in M-theory, and those who had labored on them were welcomed back into the string community. Obviously, progress would have been faster had these ideas not been excluded from consideration for so long.

There are several ideas that might help string theory solve its key problems, but they have not been widely studied. One is an old idea that the number system called *octonions* is the key to a deep understanding of the relationship between supersymmetry and higher dimensions. Another is the requirement, which I have already emphasized, that the fundamental formulation of string theory or M-theory, so far unexplored, must be background-independent. During a panel discussion on "The Next Superstring Revolution" at the 2005 Strings conference, Stephen Shenker, the director of the Stanford Institute for Theoretical Physics, observed that it was likely to come from a topic outside string theory. If this is recognized by the field's leaders, why do they not encourage young people to explore a wider range of subjects?

The narrowness of the research agenda seems to be a result of the string community's huge regard for the views of a few individuals. String theorists are the only scientists I've ever met who typically want to know what the senior people in the field, such as Edward Witten, think before expressing their own views. Of course Witten thinks clearly and deeply, but the point is that it is not good for any field if any one person's views are taken too authoritatively. There is no scientist, not even Newton or Einstein, who was not wrong on a substantial number of issues they had strong views about. Many times, in discussion after a conference talk or in conversation, if a controversial issue comes up, someone invariably asks, "Well, what does Ed think?" This used to drive me to distraction, and occasionally I would let it show: "Look, when I want to know what Ed thinks, I ask him. I'm asking you what *you* think, because I'm interested in your opinion."

Noncommutative geometry is an example of a field that was ignored by string theorists until it was embraced by Witten. Alain Connes, its inventor, tells the following story:

> I went to Chicago in 1996 and gave a talk in the Physics Department. A well-known physicist was there, and he left the room before the talk was over. I didn't meet this physicist again until two years later, when I gave the same talk in the Dirac Forum in Rutherford Laboratory, near Oxford. The same physicist was attending, this time looking very open and convinced. When he gave his talk later, he mentioned my talk quite positively. This was amazing to me, because it was the same talk, and I had not forgotten his previous reaction. So on the way back to Oxford, I was sitting next to him in the bus, and I asked him, "How can it be that you attended the same talk in Chicago and you left before the end, and now you really liked it?" The guy was not a beginner — he was in his forties. His answer was "Witten was seen reading your book in the library in Princeton!"[8]

It should be said that this attitude is fading, probably in response to the current uproar surrounding the landscape. Until last year, I had hardly ever encountered an expression of doubt from a string theorist. Now I sometimes hear from young people that there is a "crisis" in string theory. "We have lost our leaders," some of them will say. "Before this, it was always clear what the hot direction was, what people should be working on. Now there's no real guidance," or (to each other, nervously) "Is it true that Witten is no longer doing string theory?"

Another facet of string theory that many find disturbing is what can only be described as the messianic tendency of some of its practitioners, especially some younger ones. For them, string theory has become a religion. Those of us who publish papers questioning results or claims of string theorists regularly receive e-mails whose mildest form of abuse is "Are you kidding?" or "Is this a joke?" Discussions of string theory's "opponents" abound on Web sites and chat boards, where, even given the unbridled nature of such venues, the intelligence and professional competence of non–string theorists is questioned in remarkably unpleasant terms. It's hard not to conclude that at least some string theorists have begun to see themselves as crusaders rather than scientists.

Related to this swaggering is a tendency to read the evidence in the most optimistic way possible. My colleagues in quantum gravity commonly take a hard-nosed, often pessimistic, view of the prospects for solving open problems. Among loop quantum gravity theorists, I seem to be the great optimist. But my optimism pales compared with that of most string theorists. This is especially true when it comes to the big unanswered questions. As discussed, the "stringy" view of things is based on long-standing conjectures that are widely believed by string theorists but which have never been proved. Some string theorists believe them anyway. Optimism is good to a degree, but not when it results in outright misrepresentation. Unfortunately, the picture commonly offered to the general public in books and articles and TV shows — as well as to audiences of scientists — differs substantially from what a straightforward reading of the published results suggests. For example, in a review of Leonard Susskind's 2005 book, *The Cosmic Landscape,* in a trade magazine for physicists, the reviewer, reflecting on the existence of many string theories, states:

> This problem is cured by M-theory, a unique, all-embracing theory that subsumes the five superstring theories by requiring 11 space-time dimensions and incorporating higher-dimensional extended objects called branes. Among the achievements of M-theory is the first microscopic explanation for the entropy of a black hole, first predicted in the 1970s by Hawking using macroscopic arguments. . . . The problem with M-theory is that although its equations may be unique, it has billions and billions of different solutions.[9]

The most striking exaggeration here is the implication that M-theory exists as a precise theory rather than as a proposal and that it has definite equations, neither of which is true. The claim to explain black-hole entropy is (as noted in chapter 9) exaggerated, because the string theory results work only for special and atypical black holes.

You can also find such distortions in Web pages aiming to introduce string theory to the public, such as the following:

> There is even a mode describing the graviton, the particle carrying the force of gravity, which is an important reason why String Theory has received so much attention. The point is that we can make sense of the interaction of two gravitons in String theory in a way we could not in QFT. There are no infinities! And gravity is not something we put in by hand. It has to be there in a theory of strings. So, the first great achievement of String Theory was to give a consistent theory of quantum gravity.[10]

Those responsible for this particular Web page know that no one has proved there are "no infinities." But they seem to be confident enough of the truth of the conjecture to present it as fact. Further on, they, too, bring up the issue of the five different superstring theories:

> And only then it was realized that those five string theories are actually islands on the same planet, not different ones! Thus there is an underlying theory of which all string theories are only different aspects. This was called M-theory. The M might stand for Mother of all theories or Mystery, because the planet we call M-theory is still largely unexplored.

This clearly states that "there is an underlying theory," even if the last line concedes that M-theory is "still largely unexplored." A member of the public would conclude from this that there is a theory called M-theory, with the usual attributes of a theory, which is a formulation in terms of precise principles and a representation by precise equations.[11]

Many review papers and talks make equally vague and imprecise statements about results. There is unfortunately a great deal of confusion about what has actually been achieved by string theory, along with a tendency to exaggerate results and minimize difficulties. When I've queried experts, I've been shocked to find that many string theorists are unable to give correct and detailed answers to questions about the status of key conjectures, such as perturbative finiteness, S-duality, the Maldacena conjecture, or M-theory.

I understand that this a strong charge to bring, so let me illustrate it with an example. One of the basic claims made for string theory is that it is a finite theory. This means that the answers it gives to all physically sensible questions involve finite numbers. Clearly, any viable theory must provide finite answers to questions about probabilities, or finite predictions for the mass or energy of some particle or for the strength of some force. However, proposed quantum theories of fundamental forces frequently fail to do so. Indeed, of the huge number of different theories of forces consistent with the principles of relativity, all but a small number produce infinite answers to these kinds of questions. This is especially true of quantum theories of gravity. Many once promising approaches were abandoned because they failed to give finite answers. The few exceptions include string theory and loop quantum gravity.

As I discussed in chapter 12, the claim that string theory gives finite answers is couched in a certain approximation scheme called string perturbation theory. This technique gives an infinite set of approximations to the motions and interactions of strings in a given setting. We speak of the first approximation, the second approximation, the seventeenth approximation, the 100 millionth approximation, all the way up to infinity. To prove a theory finite in such a scheme, one must prove that each one of the infinite number of terms is finite. This is hard to do, but it's not impossible. It was done, for example, for the quantum theory of electromagnetism, or QED, in the late 1940s and 1950s. This was the triumph of Richard Feynman, Freeman Dyson, and their generation. The standard model of particle physics was proved finite in 1971 by Gerard 't Hooft.

The big excitement of 1984–85 was due partly to the five original superstring theories being proved finite to the first approximation. A few years later, a paper was published by the well-established theorist Stanley Mandelstam that was taken as proving all of the infinite number of terms finite.[12]

At the time, the response to Mandelstam's paper was mixed. Indeed, there is an intuitive argument — which many string theorists believe — strongly suggesting that if the theory exists at all, it will give finite answers. At the same time, several mathematicians I knew who were experts in the technical issues involved denied that the argument was a complete proof.

I didn't hear much about the issue of finiteness for many years. It simply faded into the background as the field moved on to other problems. From time to time, a paper would appear on the Internet addressing the issue, but I paid little attention. Indeed, I don't recall doubting the finiteness of the theory at all, until recently. Most of the developments I followed in the last twenty years, and a good deal of my own work in the area, were based on the assumption that string theory was finite. I heard many talks by string theorists over the years that began with the claim that the theory gave a "finite quantum theory of gravity," before going on to deal with a problem of current interest. Many books were written, and talks given for the public, asserting that string theory was a sensible quantum theory of gravity and either explicitly or implicitly claiming that the theory was finite. As far as my own work was concerned, I believed that string theory had been proved finite (or almost proved finite, up to filling in some technical detail only a mathematician would worry about), and this was a major reason for my continuing interest in it.

In 2002, I was asked to write and present a review of the whole field of quantum gravity to a conference being organized in honor of John Wheeler, one of its founders. I decided the best way to review the subject would be to write down a list of all the major results established so far by the various approaches. My hope was to make an objective comparison of how well each approach was doing in the drive toward the goal of a theory of quantum gravity. I wrote a draft of the paper and, naturally, one of the results on my list was the finiteness of superstring theory.

To finish the paper, I of course had to find proper citations to papers where each of the results listed was demonstrated. For most of them, this proved no problem, but I ran into trouble in my search for the right citation for the proof of the finiteness of string theory. Looking at different sources, I found referenced only the original paper by Mandelstam — the one that, I had been told by mathematicians, was incomplete. I found a few other papers on the problem, none of them claiming a final result. I then began asking string theorists I knew, in person and by e-mail, about the status of finiteness and where I could find the paper containing the proof. I asked a dozen or so string theorists, young and old. Almost all who answered told me that the result was true. Most didn't have the citation for the proof, and those who did gave me the paper by Mandelstam. In frustration, I consulted review papers — these are papers written to survey the main results of a field. Of more than fifteen review articles I consulted, most either said or implied that the theory was finite.[13] For citations, I found only earlier review papers or the paper of Mandelstam. I did find one review paper, by a Russian physicist, explaining that the result was unproved.[14] But it was hard to believe that he was right and all the reviews by better-known people, most of whom I knew and admired, were wrong.

Finally, I queried my Perimeter colleague Robert Myers. He told me, with his usual refreshing candor, that he didn't know whether finiteness had been completely proved, but he suggested that someone named Eric D'Hoker might. I looked him up, and this is how I finally found that D'Hoker and Phong had, just in 2001, succeeded in proving the finiteness to the second order of approximation (see chapter 12). Until then, over the seventeen years since 1984, no substantial progress had been made. (As I mentioned in chapter 12, there has been some progress in the four years since D'Hoker and Phong's paper, mainly by Nathan Berkovits. But his proof relies on additional unproved assumptions, so, while it is a step forward, it is not yet a complete proof of finiteness.) Thus, the fact was that only the first three out of an infinite number of terms in the approximation were known to be finite. Beyond that, whether the theory is finite or infinite was (and is) simply not known.

When I described this situation in my review paper, it was greeted with disbelief. I got several e-mails, not all of them polite, claiming that I was mistaken, that the theory was finite, and that Mandelstam had proved it. I had a similar experience talking to string theorists; some of them were shocked to hear that the proof of finiteness had never been completed. But their shock was as nothing compared with that of those physicists and mathematicians I talked to who were not string theorists, and who had believed that string theory was finite because they had been told that it was. For all of us, the impression of string theory as finite had had a great deal to do with our acknowledgment of its importance. None of us could re-call ever having heard a string theorist point to it as an unsolved problem.

I also felt somewhat peculiar at having to present a paper that aimed to make a detailed assessment of the evidence supporting various conjectures in string theory. Certainly, I thought, this was something that one of the leaders of the field should be doing peri-odically. This kind of critical review paper, emphasizing the key un-solved problems, is common in quantum gravity, cosmology, and, I suspect, most other fields of science. Because this was not done by any of the leaders of string theory, it was left to someone like me, as a quasi "insider" who had the technical knowledge but not the so-ciological commitment, to take on that responsibility. And I had done so because of my own interest in string theory, which I was working on almost exclusively at the time. Nevertheless, some string theorists regarded the review as a hostile act.

Carlo Rovelli, of the Centre de Physique Théorique in Marseille, is a good friend who works in quantum gravity. He had the same ex-perience when he incorporated the statement that string theory had never been proved finite into a dialogue he wrote dramatizing the debate between the different approaches to quantum gravity. He got so many e-mails asserting that Mandelstam had proved the theory finite that he decided to write to Mandelstam himself and ask his view. Mandelstam is retired, but he responded quickly. He ex-plained that what he had proved is that a certain kind of infinite term does not appear anywhere in the theory. But he told us that he had not actually proved that the theory itself was finite, because other kinds of infinite terms might appear.[15] No such term has ever been seen in any calculation done so far, but neither has anyone proved that one couldn't appear.

None of the string theorists I've discussed these issues with have decided, on learning that the theory has not been proved finite, to stop working on string theory. I've also encountered well-known string theorists who insisted that they had proved the theory's finiteness decades ago and didn't publish only because of some tech-nical issues that remained unresolved.

But when and if the issue of finiteness is settled, we will have to ask how it happened that so many members of a research program were unaware of the status of one of the key results in their field. Should it not be of concern that between 1984 and 2001 many string theorists talked and wrote as if it were a fact that the theory was finite? Why did many string theorists feel comfortable talking to outsiders and insiders alike, using language that implied the theory was fully finite and consistent?

Finiteness is not the only example in string theory of a conjecture that is widely believed but so far unproved. As we discussed, there are several versions of the Maldacena conjecture in the literature, and they have very different implications. What is sure is that the strongest of these conjectures is far from proved, although some weak version is certainly well supported. But this is not how some string theorists see it. In a recent review of the Maldacena conjecture, Gary Horowitz and Joseph Polchinski compare it to a well-known unsolved conjecture in mathematics, the Riemann hypothesis.

> In summary, we see convincing reason to place [Maldacena's duality conjecture] in the category of *true but not proven*. Indeed, we regard it on much the same footing as such mathematical conjectures as the Riemann hypothesis. Both provide unexpected connections between seemingly different structures . . . and each has resisted either proof or disproof in spite of concentrated attention.[16]

I've never heard a mathematician refer to a result as "true but not proven," but beyond that, what is astounding about this assertion is that the authors, two very smart people, ignore an obvious difference between the two cases they discuss. We know that the structures related by the Riemann hypothesis both exist mathematically; what is in question is only a conjectured relation between them. But we do not know that either string theory or the supersymmetric gauge theories really exist as mathematical structures; indeed, *their existence is part of what is in question.* What this quote makes clear is that these authors reason from the assumption that string theory is a well-defined mathematical structure — despite wide agreement that even if it is true, we have no idea what that structure is. If you don't make this unproved assumption, then your evaluation of the evidence for the strongest version of the Maldacena conjecture must disagree with theirs.

When it comes to defending their belief in these unproved conjectures, string theorists often note that something is "generally believed" in the string theory community, or that "no sensible person doubts that it's true." They seem to feel that appeal to consensus within their community is equivalent to rational argument. Here is a typical example, from the blog of a well-known string theorist:

> *Anyone who hasn't been asleep for the past 6 years knows that* quantum gravity in asymptotically anti–de Sitter space has unitary time evolution.... With the large accumulation of evidence for AdS/CFT, *I doubt there are many hold-outs left who doubt that the above statement holds*, not just in the semiclassical limit that Hawking considers, but in the full nonperturbative theory.[17] (Italics mine.)

It doesn't feel good to have to admit to being one of the hold-outs, but that is what a detailed examination of the evidence forces me to be.

This cavalier attitude toward precise support for key conjectures is counterproductive for several reasons. First, in combination with the tendencies described earlier, it means that almost no one works on these important open problems — making it more likely that they will remain unsolved. It also leads to a corrosion of the ethics and methods of science, because a large community of smart people are willing to believe key conjectures without demanding to see them proved.

Moreover, when great results are discovered, they are often exaggerated. Several non–string theorists have asked me why I work on anything else, when string theory has completely explained black-hole entropy. While I greatly admire the work on extremal black holes by Strominger and Vafa and others (see chapter 9), I must reiterate that, for what appear to be good reasons, the precise results have not been extended to black holes in general.

Similarly, the claim that a vast number of string theories exist with a positive cosmological constant (the much-discussed "landscape") is far from secure. Yet some leading string theorists are willing, on the basis of these weak results, to make grand pronouncements about string theory's success and future prospects.

It may well be that the persistent exaggeration has benefited string theory over its rivals. If you were a department head or an officer of a granting agency, wouldn't you be more likely to fund or hire a scientist who worked on a program said to solve the big problems in the field over a scientist who could claim only that he or she had evidence that there might exist a theory — so far unformulated — that had the potential to solve the problems?

Let me summarize, so we can see where this is taking us. The discussion has brought out seven unusual aspects of the string theory community.

1. *Tremendous self-confidence,* leading to a sense of entitlement and of belonging to an elite community of experts.

2. *An unusually monolithic community,* with a strong sense of consensus, whether driven by the evidence or not, and an unusual uniformity of views on open questions. These views seem related to the existence of a hierarchical structure in which the ideas of a few leaders dictate the viewpoint, strategy, and direction of the field.

3. In some cases, a *sense of identification with the group,* akin to identification with a religious faith or political platform.

4. A strong scnsc of *the boundary between the group and other experts.*

5. A *disregard for and disinterest in* the ideas, opinions, and work of experts who are not part of the group, and a preference for talking only with other members of the community.

6. A tendency to *interpret evidence optimistically,* to believe exaggerated or incorrect statements of results, and to disregard the possibility that the theory might be wrong. This is coupled with a tendency to *believe results are true because they are "widely believed,"* even if one has not checked (or even seen) the proof oneself.

7. A lack of appreciation for the extent to which a research program ought to involve risk.

Of course, not all string theorists can be described this way, but few observers, inside or outside the string theory community, will disagree that some or all of these attitudes characterize that community.

I want to be clear that I am not criticizing the behavior of specific individuals. Many string theorists are personally open-minded and self-critical, and if asked, they will say that they deplore these characteristics of their community.

I must also be clear that I am as much at fault as my colleagues in string theory. For many years, I believed that basic conjectures such as finiteness were proved. This is largely why I invested years of work in string theory. More than just my own work was affected, for among the community of people who work on quantum gravity, I was the strongest advocate for taking string theory seriously. Yet I did not take the time to check the literature, so I, too, was willing to let the leaders of the string theory community do my critical thinking for me. And during the years I worked on string theory, I cared very much what the leaders of the community thought of my work. Just like an adolescent, I wanted to be accepted by those who were the most influential in my little circle. If I didn't actually take their advice and devote my life to the theory, it's only because I have a stubborn streak that usually wins out in these situations. For me, this is not an issue of "us" versus "them," or a struggle between two communities for dominance. These are very personal problems, which I have been contending with internally for as long as I have been a scientist.

So I sympathize strongly with the plight of string theorists, who want both to be good scientists and to have the approval of the powerful people in their field. I understand the difficulty of thinking clearly and independently when acceptance in your community requires belief in a complicated set of ideas that you don't know how to prove yourself. This is a trap it took me years to think my way out of.

All of which bolsters my conviction that we theoretical physicists are in trouble. If you ask many string theorists why scientists working on alternatives to string theory are never invited to string theory conferences, they will agree with you that such people should be invited, they will deplore the current state of affairs, but they will insist that there's nothing they can do about it. If you ask them why string theory groups never hire young people working on alternatives as postdocs or faculty or invite them as visitors, they will agree with you that this would be a good thing to do, and they will lament the fact that it isn't being done. The situation is one in which there are big issues that many agree on but no one feels responsible for.

I strongly believe in my string theory friends. I believe that as individuals, they are almost all more open-minded and self-critical and less dogmatic than they are *en masse*.

How could a community act in a way so at odds with the good-will and good sense of its individual members?

It turns out that sociologists have no problem recognizing this phenomenon. It afflicts communities of highly credentialed experts, who by choice or circumstance communicate only among themselves. It has been studied in the context of intelligence agencies and governmental policy-making bodies and major corporations. Because the consequences have sometimes been tragic, there is a literature describing the phenomenon, which is called *groupthink*.

Yale psychologist Irving Janis, who coined the term in the 1970s, defines groupthink as "a mode of thinking that people engage in when they are deeply involved in a cohesive in-group, when the members' strivings for unanimity override their motivation to realistically appraise alternative courses of action."[18] According to this definition, groupthink occurs only when cohesiveness is high. It requires that members share a strong "we-feeling" of solidarity and a desire to maintain relationships within the group at all costs. When colleagues operate in a groupthink mode, they automatically apply the "preserve group harmony" test to every decision they face.[19]

Janis was studying failures of decision making by groups of experts, such as the Bay of Pigs. The term has since been applied to many other examples, including the failure of NASA to prevent the *Challenger* disaster, the failure of the West to anticipate the collapse of the Soviet Union, the failure of the American automobile companies to foresee the demand for smaller cars, and most recently — and perhaps most calamitously — the Bush administration's rush to war on the basis of a false belief that Iraq had weapons of mass destruction.

Here's a description of groupthink excerpted from an Oregon State University Web site dealing with communication:

Groupthink members see themselves as part of an in-group working against an outgroup opposed to their goals. You can tell if a group suffers from groupthink if it:

    1. overestimates its invulnerability or high moral stance,

    2. collectively rationalizes the decisions it makes,

    3. demonizes or stereotypes outgroups and their leaders,

    4. has a culture of uniformity where individuals censor themselves and others so that the facade of group unanimity is maintained, and

    5. contains members who take it upon themselves to protect the group leader by keeping information, theirs or other group members', from the leader.[20]

This does not match up one-to-one with my characterization of the culture of string theory, but it's close enough to be worrying.

Of course, string theorists will not have any trouble answering this critique. They can point to many historical examples showing that the progress of science depends on the establishment of tight consensus among a community of experts. The views of outsiders must be disregarded because outsiders are not skilled enough in the tools of the trade to evaluate evidence and pass judgment. It follows that the scientific community must have mechanisms for the establishment and enforcement of consensus. What may seem like groupthink to an outsider is in fact rationality, displayed according to tightly constrained rules.

They can also counter the charge that they let the consensus of their research community substitute for the critical thinking of individuals. According to one prominent sociologist of science I've discussed this with, the fact that key conjectures are believed without being proved is not unusual.[21] No scientist can directly confirm more than a small fraction of the experimental results, calculations, and proofs that form the foundations of their beliefs about their subject; few have the skills, and in contemporary science no one has the time. Thus, when you join a scientific community, you must trust your colleagues to tell the truth about the results in their domains of expertise. This can lead to a conjecture being accepted as a fact, but it happens as often in research programs that are ultimately successful as it does in those that fail. Contemporary science just cannot be done without a community of people who trust what members tell them. Thus, while episodes like this are to be regretted and should be corrected when found, they are not by themselves indications of a doomed research program or a pathological sociology.

Finally, senior string theorists can argue that they deserve their credentials and with those comes the right to direct research as they see fit. After all, the practice of science is based on hunches, and this is their hunch. Would someone have them waste their time working on something they don't believe in? And should they hire people who work on theories other than the one they believe has the best chance of success?

But how does one respond to this defense? If science is based on consensus among a community of experts, then what you have in string theory is a community of experts who are in remarkable agreement about the ultimate correctness of the theory they study. Is there any rational ground to stand on — any way to mount an intelligent and useful dissent? We need to do much more than throw around terms like "groupthink." We must have a theory of what science is and how it works, one that clearly demonstrates why it is bad for science when a particular community comes to dominate a field before its theory has passed the usual tests of proof. This is the task to which we now turn.

17

What Is Science?

To reverse the troubling trends in physics, we must first understand what science is — what moves it forward and what holds it back. And to do this, we must define science as something more than the sum of what scientists do. The aim of this chapter is to propose such a definition.

When I entered graduate school at Harvard in 1976, I was a naïve student from a small college. I was in awe of Einstein, Bohr, Heisenberg, and Schrödinger and how they had changed physics through the force of their radical thinking. I dreamed, as young people do, of being one of them. I now found myself at the center of particle physics, surrounded by the leaders in the field — people like Sidney Coleman, Sheldon Glashow, and Steven Weinberg. These people were incredibly smart, but they were nothing like my heroes. In lectures, I never heard them talk about the nature of space and time or issues in the foundations of quantum mechanics. Neither did I meet many students with these interests.

This led me to a personal crisis. I was certainly not as well prepared as students from the great universities, but I had done research as an undergraduate, which most of my peers had not, and I knew I was a quick study. So I was confident that I could do the work. But I also had a very particular idea of what a great theoretical physicist should be. The great theoretical physicists I was rubbing shoulders with at Harvard were rather different from that. The atmosphere was not philosophical; it was harsh and aggressive, dominated by people who were brash, cocky, confident, and in some cases insulting to people who disagreed with them.

During this time, I made friends with a young philosopher of science named Amelia Rechel-Cohn. Through her, I came to know people who, like me, were interested in the deep philosophical and foundational issues in physics. But this only made matters worse. They were nicer than the theoretical physicists, but they seemed happy just to analyze precise logical issues in the foundations of special relativity or ordinary quantum physics. I had little patience for such talk; I wanted to *invent* theories, not criticize them, and I was sure that — as unreflective as the originators of the standard model seemed — they knew the things I needed to know if I was to get anywhere.

Just as I started to think seriously about quitting, Amelia gave me a book by the philosopher Paul Feyerabend. It was called *Against Method* and it spoke to me — but what it had to say was not very encouraging. It was a blow to my naïveté and self-absorption.

What Feyerabend's book said to me was: *Look, kid, stop dreaming! Science is not philosophers sitting in clouds. It is a human activity, as complex and problematic as any other. There is no single method to science and no single criterion for who is a good scientist. Good science is whatever works at a particular moment of history to advance our knowledge. And don't bother me with how to define progress — define it any way you like and this is still true.*

From Feyerabend, I learned that progress sometimes requires deep philosophical thinking, but most often it does not. It is mostly furthered by opportunistic people who cut corners, exaggerating what they know and have accomplished. Galileo was one of these; many of his arguments were wrong, and his opponents — the well-educated, philosophically reflective Jesuit astronomers of the time — easily punched holes in his thinking. Nevertheless, he was right and they were wrong.

What I also learned from Feyerabend is that no a priori argument can tell us what will work in all circumstances. What works to advance science at one moment will be wrong at another. And I learned one more thing from his stories of Galileo: You have to fight for what you believe.

Feyerabend's message was a none too timely wake-up call. If I wanted to do good science, I had to recognize that the people I was lucky enough to be studying with were indeed the great scientists of the day. Like all great scientists, they had succeeded because their ideas were right and they had fought for them. If your ideas are right and you fight for them, you'll accomplish something. Don't waste time feeling sorry for yourself or waxing nostalgic about Einstein and Bohr. No one but you can develop your ideas, and no one but you will fight for them.

I took a long walk and decided to stay in science. I soon found I could do real research by applying the methods used by particle physicists to the problem of quantum gravity. If this meant leaving aside, for a time, the foundational issues, it was nevertheless wonderful to be able to invent new formulations and do some calculations within them.

To thank him for saving my career, I sent a copy of my PhD thesis to Feyerabend. In reply, he sent me his new book, *Science in a Free Society* (1979), with a note inviting me to look him up if I was ever in Berkeley. A few months later, I happened to be in California for a particle-physics conference and tried to track him down, but it was quite a project. He kept no office hours at the university and, indeed, no office. The Philosophy Department secretary laughed when I asked for him and advised me to try him at home. There he was in the phone book, on Miller Avenue in the Berkeley hills. I summoned up my courage, dialed, and politely asked for Professor Paul Feyerabend. Whoever was on the other end shouted *"Professor Paul Feyerabend! That's the other Paul Feyerabend. You can find him at the university"* and hung up. So I dropped in on one of his classes, and found him happy to talk afterward, if only briefly. But in the few minutes he gave me, he offered an invaluable piece of advice. "Yes, the academic world is screwed up, and there's nothing you can do about it. But don't worry about that. Just do what you want. If you know what you want to do and advocate for it, no one will put any energy into stopping you."

Six months later, he wrote me a second note, which reached me in Santa Barbara, where I had just accepted a position as a postdoc at the Institute for Theoretical Physics. He mentioned that he was talking with a talented physics undergraduate who, like me, had philosophical interests. Would I like to meet him and advise him on how to proceed? What I really wanted was another chance to talk to Feyerabend, so I went up to Berkeley again and met the two of them on the steps of the philosophy building (as close, apparently, as he ever got to his colleagues.) Feyerabend treated us to lunch at Chez Panisse, then took us up to his house (which turned out to be on Miller Avenue, in the Berkeley hills) so the student and I could talk while he watched his favorite soap opera. On the way, I shared the backseat of Feyerabend's little sports car with the inflatable raft he kept there in case an 8-point earthquake came while he was on the Bay Bridge.

The first topic Feyerabend brought up was renormalization, the method for dealing with infinities in quantum field theory. I was surprised to find that he knew quite a lot of contemporary physics. He wasn't antiscience, as some of my professors at Harvard had intimated he would be. It was clear that he loved physics, and he was more conversant with the technicalities than most philosophers I had met. His reputation as hostile to science had undoubtedly arisen because he considered the question of why science worked as unanswered. Was it because science has a method? So do witch doctors.

Perhaps the difference, I ventured, is that science uses math. And so does astrology, he responded, and he would have explained the details of the various computational systems used by astrologers, if we had let him. Neither one of us knew what to say when he argued that Johannes Kepler, one of the greatest physicists who ever lived, had made several contributions to the technical refinement of astrology, and Newton had spent more time on alchemy than on physics. Did we think we were better scientists than Kepler or Newton?

Feyerabend was convinced that science is a human activity, carried out by opportunistic people who follow no general logic or method and who do whatever it takes to increase knowledge (however you define it). So his big question was: How does science work, and why does it work so well? Even though he countered all of my own explanations, I sensed that he passionately pursued the question not because he was antiscience but because he cared about it.

As the day progressed, Feyerabend told us his story. He had been a physics prodigy as a teenager in Vienna, but his studies were curtailed when he was drafted to fight in World War II. He was wounded on the Russian front and later ended up in Berlin, where he found work after the war as an actor. After a time, he tired of the theater world and returned to the study of physics in Vienna. He joined the philosophy club, where he discovered that he could win on any side of a philosophical debate simply by using the skills he had learned in the acting profession. This made him wonder whether academic success had any rational basis. One day, the students succeeded in inviting Ludwig Wittgenstein to come to their club. Feyerabend was so impressed that he decided to switch to philosophy. He spoke with Wittgenstein, who invited him to come to Cambridge to study with him. But by the time Feyerabend got to England, Wittgenstein was dead, so someone suggested he talk with Karl Popper, another Viennese expatriate, who was teaching at the London School of Economics. So he moved to London and began his life in philosophy by writing papers attacking Popper's work.

After a few years, he was offered a teaching job. He asked a friend how he could possibly teach, given how little he knew. The friend told him to write down what he thought he knew. It filled a single piece of paper. The friend then told him to make the first sentence the subject of the first lecture, the second sentence the subject of the second lecture, and so on. So the physics student turned soldier turned actor became a philosophy professor.[1]

Feyerabend drove us back down to the Berkeley campus. Before leaving us, he gave us a last piece of advice. "Just do what you want to do and don't pay any attention to anything else. Never in my career have I spent five minutes doing something I didn't want to be doing."

And this is more or less what I have done. Until now. Now I feel we have to talk not just about scientific ideas but also about the scientific process. There is no choice. We have a responsibility to the generations that follow to think about why we have been so much less successful than our teachers.

Since my visit with Feyerabend, who died in 1994 at the age of seventy, I have mentored several talented young people through crises very similar to my own. But I cannot tell them what I told my younger self — that the dominant style was so dramatically success-ful that it must be respected and accommodated. Now I have to agree with my younger colleagues that the dominant style is not succeeding.

First and foremost, the style of doing science that I learned at Harvard has not led to more progress. It succeeded in establishing the standard model but failed to go beyond it. After thirty years, we must ask whether that style has, for the time being, outlived its use-fulness. Perhaps this is a moment that requires the more reflective, risky, and philosophical style of Einstein and his friends.

The problem is far wider than string theory; it involves the val-ues and attitudes fostered by the physics community as a whole. Put simply, the physics community is structured in such a way that large research programs that promote themselves aggressively have an advantage over smaller programs that make more cautious claims. Therefore, young academic scientists have the best chance of succeeding if they impress older scientists with technically sweet solutions to long-standing problems posed by dominant research programs. To do the opposite — to think deeply and independently and try to formulate one's own ideas — is a poor strategy for success.

Physics thus finds itself unable to solve its key problems. It is time to reverse course — to encourage small, risky new research programs and discourage the entrenched approaches. We ought to be giving the advantage to the Einsteins — people who think for them-selves and ignore the established ideas of powerful senior scientists.

But to convince the skeptics, we have to answer Feyerabend's question about how science works.

There seem to be two conflicting views of science. One is science as the domain of the rebel, the individual who comes up with grand new ideas and works hard over a lifetime to prove them true. This is the myth of Galileo, and we see it played out today in the efforts of a few greatly admired scientists, like the mathematical physicist Roger Penrose, the complexity theorist Stuart Kauffman, and the bi-ologist Lynn Margulis. Then there is the view of science as a conser-vative, consensual community that tolerates little deviation from orthodox thinking and channels creative energy into furthering well-defined research programs.

In some sense, both views are true. Science requires both the rebel and the conservative. This seems at first paradoxical. How has an enterprise flourished for centuries that requires the conservative and the rebel to coexist? The trick seems to be to bring the rebel and conservative into lifelong and uncomfortable proximity, within the community and, to some extent, within each individual as well. But how is this accomplished?

Science is a democracy, in that every scientist has a voice, but it is nothing like majority rule. Still, whereas individual judgment is prized, consensus plays a crucial role. Indeed, what ground can I stand on when the majority of my profession embraces a research program I cannot accept even though accepting it would be to my benefit? The answer is that democracy is much more than rule by the majority. There is a system of ideals and ethics that transcends majority rule.

Thus, if we are to argue that science is more than sociology, more than academic politics, we must have a notion of what science is that is consistent with, but more than, the idea of a self-governing community of human beings. To argue that a particular form of organization, a particular behavior, is good or bad for science, we must have a basis for making value judgments that goes beyond what is popular. We must have a basis for disagreeing with the majority without being labeled a crank.

Let us start by breaking Feyerabend's question down into a few simpler questions. We can say that science progresses when scientists reach consensus on a question. What are the mechanisms that govern how this happens? Before a consensus is reached, there is often controversy. What is the role of disagreement in preparing the way for scientific progress?

To answer these questions, we should go back to the views of earlier philosophers. In the 1920s and 1930s, a philosophical movement grew up in Vienna called *logical positivism*. The logical positivists proposed that assertions become knowledge when they are verified by observations of the world, and they claimed that scientific knowledge is the sum of these verified propositions. Science progresses when scientists make assertions that have verifiable content, which are then verified. Their motive was to rid philosophy of metaphysics, which had filled huge books with statements that made no contact with reality. In this they partly succeeded, but their modest characterization of science did not last. There were many problems, one of which was that there is no ironclad correspondence between what is observed and what is stated. Assumptions and biases creep into the descriptions of the simplest observations. It is not practical, perhaps not even possible, to break what scientists say or write into little atoms, each of which corresponds to an observation stripped of theorizing.

When verificationism failed, philosophers proposed that science progresses because scientists follow a method guaranteed to lead to the truth. Proposals for the scientific method were offered by philosophers such as Rudolf Carnap and Paul Oppenheim. Karl Popper put forward his own proposal, which was that science progresses when scientists propose theories that are falsifiable — that is, they make statements that can be contradicted by experiment. According to Popper, a theory is never proved right, but if it survives many attempts to prove it wrong, we can begin to have faith in it — at least until it is finally falsified.[2]

Feyerabend began his work in philosophy by attacking these ideas. For example, he showed that falsifying a theory is not such an easy thing. Very often, scientists keep a theory alive after it appears to have been falsified; they do this by changing the interpretation of the experiment. Or they challenge the results themselves. Sometimes this leads to a dead end because the theory really is wrong. But sometimes keeping a theory alive in the face of apparent experimental contradiction turns out to be the right thing to do. How can you tell which situation you're in? Feyerabend argued that you can't. Different scientists adopt different viewpoints and take their chances on which one will be borne out by developments. There is no general rule for when to abandon a theory and when to keep it alive.

Feyerabend also attacked the whole idea that method is the key to scientific progress, by showing that at critical junctures scientists will make progress by breaking the rules. Moreover, he argued — convincingly, in my view — that science would grind to a halt were the "method's" rules always followed. The science historian Thomas Kuhn made another attack on the notion of a "scientific method" when he argued that scientists follow different methods at different times. But he was less radical than Feyerabend; he tried to set out two methods, that of "normal science" and that of scientific revolutions.[3]

Another criticism of Popper's ideas was made by the Hungarian philosopher Imre Lakatos, who argued that there was not as much asymmetry between falsification and verification as Popper supposed. If you see one bright red swan, you are not likely to give up a theory that says that all swans are white; you will instead go looking for the person who painted it.[4]

These arguments leave us with several problems. The first is that the success of science still requires explanation; the second (emphasized by Popper) is that it becomes impossible to distinguish sciences like physics and biology from other belief systems — such as Marxism, witchcraft, and intelligent design — that claim to be scientific.[5] If no such distinction can be made, the door is left open to a scary kind of relativism, in which all claims to truth and reality have equal footing.

While I'm convinced, like many practicing scientists, that we follow no single method, I'm also convinced that we must answer Feyerabend's question. We can begin by discussing the role science has played in human culture.

Science is one of several instruments of human culture that arose in response to the situation we humans have found ourselves in since prehistoric times: We, who can dream of infinite time and space, of the infinitely beautiful and the infinitely good, find ourselves embedded in several worlds: the physical world, the social world, the imaginative world, and the spiritual world. It's a condition of being human that we have long sought to discover crafts that give us power over these diverse worlds. These crafts are now called science, politics, art, and religion. Now, as in our earliest days, they give us power over our lives and form the basis of our hopes.

Whatever they have been called, there has never been a human society without science, politics, art, and religion. In caves whose walls are adorned with the paintings of ancient hunters, we have found bones and rocks with patterns showing that people were counting something in groups of fourteen, twenty-eight, or twenty-nine. The archaeologist Alexander Marshack, author of *The Roots of Civilization*, has interpreted these as observations of the phases of the moon.[6] They might also have been records of an early method of birth control. In either case, they show that twenty thousand years ago human beings were using mathematics to organize and conceptualize their experience of nature.

Science was not invented. It evolved over time, as people discovered tools and habits that worked to bring the physical world within the sphere of our understanding. Science, then, is the way it is because of the way nature is — and because of the way we are. Many philosophers mistakenly look for an explanation of why science works that would apply to any possible world. But there can be no such thing. A method that would work in any possible universe would be like a chair that would be comfortable for any possible animal: It would fit equally badly in most cases.

Indeed, it is possible to prove a version of this statement. Suppose that scientists are like blind explorers looking for the highest peak in their country. They cannot see, but they can feel around them to determine which way is up and which is down, and they have an altimeter with an audio readout to determine how high they are. They cannot see when they are on top of a peak, but they will know it because only there do all directions lead down. The problem is that there can be more than one peak, and if you can't see, it's hard to be sure you're climbing the highest one. It is thus not obvious whether there is a strategy the blind explorers can follow to find the highest peak in the least amount of time. This is a problem that mathematicians used to study, until it was proved impossible. The no-free-lunch theorem, developed by David Wolpert and William Macready, states that no strategy will do better in every possible landscape than simply moving around randomly.[7] To fashion a strategy that does better, you have to know something about your landscape. The kind of strategy that would work well in Nepal would fail in Holland.

It is thus no surprise that philosophers were unable to discover a general strategy that would explain how science works. And the strategies they did invent didn't bear much resemblance to what scientists actually do. The successful strategies were discovered over time and are embedded in the practices of the individual sciences.

Once we understand this, we can identify the features of nature that science exploits. The most important is that nature is relatively stable. In physics and chemistry, it's easy to devise experiments whose results are repeatable. This did not have to be the case; for example, it is less the case in biology and far less in psychology. But in the domains where experiments are repeatable, it is useful to describe nature in terms of laws. Thus, from its beginnings, the practitioners of physics have been interested in discovering general laws. What is at issue here is not whether there actually are fundamental laws; what matters for how we do science is whether there are regularities that we can discover and model, using tools that we can make with our hands.

We happen to live in a world hospitable to our understanding, and this has always been the case. From the very beginning of our life as a species, we could easily observe regularities in the sky and in the seasons, in the migrations of animals and the growth of plants, and in our own biological cycles. By making marks on bone and rocks, we learned that we could keep track of these regularities, correlate them, and use this knowledge to our advantage. Down to the present experiments with huge telescopes, powerful microscopes, and bigger and bigger accelerators, we are doing only what we have always done: using the technology at hand to discover patterns unfolding before us.

But if science works because we live in a world of regularities, it works in the particular way it does because of some peculiarities in our own makeup. In particular, *we are masters at drawing conclusions from incomplete information.* We are constantly observing the world and then making predictions and drawing conclusions about it. That is what hunter-gatherers do, and it is also what particle physicists and microbiologists do. We never have enough information to completely justify the conclusions we draw. Being able to act on guesses and hunches, and act confidently when the information we have points somewhere but does not constitute a proof, is an essential skill that makes someone a good businessperson, a good hunter or farmer, or a good scientist. It is a big part of what makes human beings such a successful species.

But this ability comes at a heavy price, which is that we easily fool ourselves. Of course, we know that we're easily fooled by others. Lying is strongly sanctioned because it is so effective. It is, after all, only because we are built to come to conclusions from incomplete information that we are so vulnerable to lies. Our basic stance has to be one of trust, for if we required proof of everything, we would never believe anything. We would then never *do* anything — never get out of bed, never make marriages, friendships, or alliances. Without the ability to trust, we would be solitary animals. Language is effective and useful because most of the time we believe what other people tell us.

But what is equally important, and sobering, is how often we fool ourselves. And we fool ourselves not only individually but *en masse.* The tendency of a group of human beings to quickly come to believe something that its individual members will later see as obviously false is truly amazing. Some of the worst tragedies of the last century happened because well-meaning people fell for easy solutions proposed by bad leaders. But arriving at a consensus is part of who we are, for it is essential if a band of hunters is to succeed or a tribe is to flee approaching danger.

For a community to survive, then, there must be mechanisms of correction: elders who curb the impulsiveness of the young because if they have learned anything from their long lives, it is how often they were wrong; the young, who challenge beliefs that have been held obvious and sacred for generations, when those beliefs are no longer apt. Human society has progressed because it has learned to require of its members both rebellion and respect, and because it has discovered social mechanisms that over time balance those qualities.

I believe that science is one of those mechanisms. It is a way to nurture and encourage the discovery of new knowledge, but more than anything else it is a collection of crafts and practices that, over time, have been shown to be effective in unmasking error. It is our best tool in the constant struggle to overcome our built-in tendency to fool ourselves and fool others.

From this brief sketch, we can see what science and the democratic process have in common. Both the scientific community and the community at large need to reach conclusions and make decisions based on incomplete information. In both cases, the incompleteness of information will lead to the forming of factions that hold different points of view. Societies, scientific and otherwise, need mechanisms to resolve disputes and reconcile differences of opinion. Such mechanisms require that errors be uncovered and new solutions to intractable problems be allowed to replace older ones. There are many such mechanisms in human societies, some of them involving force or coercion. The most basic idea of democracy is that a society will function best when disputes are resolved peacefully. Science and democracy, then, share a common and tragic awareness of our tendency to fool ourselves, and also the optimistic belief that as a society we can practice correctives that make us collectively, over time, wiser than any individual.

Now that we've put science in its proper context, we can turn to the question of why it works so well. I believe the answer is simple: *Science has succeeded because scientists comprise a community that is defined and maintained by adherence to a shared ethic.* It is adherence to an ethic, not adherence to any particular fact or theory, that I believe serves as the fundamental corrective within the scientific community.

There are two tenets of this ethic:

1. If an issue can be decided by people of good faith, applying rational argument to publicly available evidence, then it must be regarded as so decided.
2. If, on the other hand, rational argument from the publicly available evidence does not succeed in bringing people of good faith to agreement on an issue, society must allow and even encourage people to draw diverse conclusions.

I believe that science succeeds because scientists adhere, if imperfectly, to these two principles. To see whether this is true, let us look at some of the things these principles require us to do.

- We agree to argue rationally, and in good faith, from shared evidence, to whatever degree of shared conclusions are warranted.
- Each individual scientist is free to develop his or her own conclusions from the evidence. But each scientist is also required to put forward arguments for those conclusions for the consideration of the whole community. These arguments must be rational and based on evidence available to all members. The evidence, the means by which the evidence was obtained, and the logic of the arguments used to deduce conclusions from the

evidence must be shared and open to examination by all members.

- The ability of scientists to deduce reliable conclusions from the shared evidence is based on the mastery of tools and procedures developed over many years. They are taught because experience has shown that they often lead to reliable results. Every scientist trained in such a craft is deeply aware of the capacity for error and self-delusion.

- At the same time, each member of the scientific community recognizes that the eventual goal is to establish consensus. A consensus may emerge quickly, or it may take some time. The ultimate judges of scientific work are future members of the community, at a time sufficiently far in the future that they can better evaluate the evidence objectively. While a scientific program may temporarily succeed in gathering adherents, no program, claim, or point of view can succeed in the long run unless it produces sufficient evidence to persuade the skeptics.

- Membership in the community of science is open to any human being. Considerations of status, age, gender, or any other personal characteristic may not play a role in the consideration of a scientist's evidence and arguments, and may not limit a member's access to the means of dissemination of evidence, argument, and information. Entry to the community is, however, based on two criteria. The first is the mastery of at least one of the crafts of a scientific subfield to the point where you can independently produce work judged by other members to be of high quality. The second criterion is allegiance and continued adherence to the shared ethic.

- While orthodoxies may become established temporarily in a given subfield, the community recognizes that contrary opinions and research programs are necessary for the community's continued health.

When people join a scientific community, they give up certain childish but universal desires: the need to feel that they are right all the time or the belief that they are in possession of the absolute truth. In exchange, they receive membership in an ongoing enterprise that over time will achieve what no individual could ever achieve alone. They also receive expert training in a craft, and in most cases learn much more than they ever could on their own. Then, in exchange for their labor expended in the practice of that craft, the community safeguards a member's right to advocate any view or research program he or she feels is supported by the evidence developed from its practice.

I would call this kind of community, in which membership is defined by adherence to a code of ethics and the practice of crafts developed to realize them, an *ethical community*. Science, I would propose, is the purest example we have of such a community.

But it is not sufficient to characterize science as an ethical community, because some ethical communities exist to preserve old knowledge rather than to discover new truths. Religious communities, in many cases, satisfy the criteria for being ethical communities. Indeed, science in its modern form evolved from monasteries and theological schools — ethical communities whose aim was the preservation of religious dogma. So if our characterization of science is to have teeth, we must add some criteria that cleanly distinguish a physics department from a monastery.

To do this, I would like to introduce a second notion, which I call an *imaginative community*. This is a community *whose ethic and organization incorporates a belief in the inevitability of progress and an openness to the future.* The openness leaves room, imaginatively and institutionally, for novelty and surprise. Not only is there a belief that the future will be better, there is an understanding that we cannot forecast how that better future will be reached.

Neither a Marxist state nor a fundamentalist religious state is an imaginative community. They may look forward to a better future, but they believe they know exactly how that future will be reached. In words I heard often from my Marxist grandmother and friends while growing up, they are sure they are right because their "science" teaches them "the correct analysis of the situation."

An imaginative community believes that the future will bring surprises, in the form of new discoveries and new crises to be overcome. Rather than placing faith in their present knowledge, its members invest their hopes and expectations for the future in future generations, by passing along to them the ethical precepts and tools of thinking, individual and collective, that will enable them to overcome and take advantage of circumstances that are beyond the present powers of imagination.

Good scientists expect that their students will exceed them. Although the academic system gives a successful scientist many reasons to believe in his or her own authority, any good scientist knows that the minute you succumb to believing that you know more than your best students, you cease to be a scientist.

The scientific community is thus both an ethical and an imaginative community.

What should be abundantly clear from this description is that controversy is essential for the progress of science. My first principle says that when we are forced to reach a consensus by the evidence, we should do so. But my second principle says that until the evidence forces consensus, we should encourage a wide diversity of viewpoints. This is good for science — a point that Feyerabend made often, and I believe correctly. Science proceeds fastest when there are competing theories. The older, naïve view is that theories are put forward one at a time and tested against the data. This fails to take into account the extent to which the theoretical ideas we have influence which experiments we do and how we interpret them. If only one theory is contemplated at a time, we are likely to get stuck in intellectual traps created by that theory. The only way out is if different theories compete to explain the same evidence.

Feyerabend argued that even in cases where there is a widely accepted theory that agrees with all the evidence, it is still necessary to invent competing theories in order for science to progress. This is because experiments that contradict the established view are most likely to be suggested by a competing theory and perhaps would not even have been conceived were there not a competing theory. So competing theories give rise to experimental anomalies as often as the reverse.

Therefore Feyerabend insisted that scientists should never agree, unless they are forced to. When scientists come to agreement too soon, before they are compelled to by the evidence, science is in danger. We then have to ask what influenced them to come to the premature conclusion. As they are only human, the answer to this will likely be the same factors that cause people to agree about all sorts of things that don't rely on evidence, from religious beliefs to fashion to trends in popular culture.

So the question is this: Do we want scientists to come to agreement because they want to be liked or seen as brilliant by other scientists, or because everyone they know thinks the same thing, or because they like to be on the winning team? Most people are tempted to agree with other people for motives such as these. There is no reason that scientists should be immune from them, being human, after all.

Still, we must fight those urges if we want to keep science alive. We must encourage the opposite, which is to disagree as much as the evidence permits. Given how much humans need to be liked, to fit in, to be part of the winning team, we must make it clear that when we succumb to these needs, we are letting science down.

There are other reasons that a healthy scientific community should encourage disagreement. Science moves forward when we are forced to agree with something unexpected. If we think we *know* the answer, we will try to make every result fit that preconceived idea. It is controversy that keeps science alive, keeps it moving. In an atmosphere filled with controversy over rival views, sociological forces are not enough to bring people into agreement. So on those rare occasions when we do come to consensus on something, it is because we have no choice. The evidence forces us to do so, even if we don't like it. That is why progress in science is real.

There are several obvious objections to this characterization of science. First, there are clearly violations by members of the community of the ethic I have just described. Scientists often do exaggerate and distort the evidence. Age, status, fashion, peer pressure, all do play a role in the workings of the scientific community. Some research programs do succeed in gathering adherents and resources beyond what the evidence supports, while other research programs that ultimately bear fruit are suppressed by sociological forces.

But I would suggest that enough scientists adhere to enough of the ethic that in the long run progress continues to be made, despite the fact that time and resources are wasted in the promotion and defense of orthodox and fashionable ideas that later turn out to be wrong. The role of time must be emphasized. Whatever may happen in the short run, over decades evidence almost always accumulates that settles contrary claims by consensus, without regard to fashion.

Another possible objection is that the characterization I've given is logically incomplete. I do not offer any criteria for determining which crafts are necessary to master. But I think this is best done by the communities themselves, over many generations. There is no way that Newton or Darwin could have predicted the range of tools and procedures used now.

Adherence to the shared ethic is never perfect, so there is always room for improvement in the practice of science. This seems especially true today, when fashion appears to be playing too large a role, at least in physics. You know this is happening whenever there are bright young recent PhDs who tell you privately that they would rather be doing X but are doing Y because that is the direction or technique championed by powerful older people, and they thus feel the need to do Y to get funding or a job. Of course, in science as in other areas, there are always a few who choose to do X in spite of the clear evidence that the doers of Y are better rewarded in the short term. Among them are the people who will most likely lead the next generation. Thus the progress of science may be slowed by orthodoxy and fashion, but as long as there is room for those who do X instead of Y, it cannot be stopped completely.

All this is to say that like everything else that human beings do, success in science is to a large extent driven by courage and character. While the progress of science relies on the possibility of achieving consensus in the long term, the decisions an individual scientist makes as to what to do, and how to evaluate the evidence, are always based on incomplete information. Science progresses because it is built on an ethic recognizing that in the face of incomplete information we are all equal. No one can predict with certainty whether an approach will lead to definite progress or years of wasted work. All we can do is train students in the crafts that experience has shown to lead most often to reliable conclusions. After that, we must leave them free to follow their own hunches and we must make time to listen to them when they report back. As long as the community continually opens up opportunities for new ideas and points of view and adheres to the ethic that in the end we require consensus based on rational argument from evidence available to all, science will eventually succeed.

The task of forming the community of science will never be finished. It will always be necessary to fight off the dominance of orthodoxy, fashion, age, and status. There will always be temptations to take the easy way, to sign up with the team that seems to be winning rather than try to understand a problem afresh. At its finest, the scientific community takes advantage of our best impulses and desires while protecting us from our worst. The community works in part by harnessing the arrogance and ambition we each in some degree bring to the search. Richard Feynman may have said it best: *Science is the organized skepticism in the reliability of expert opinion.*[8]

18

Seers and Craftspeople

PERHAPS THERE IS something wrong with the way we are going about trying to make a revolution in physics. I argued in chapter 17 that science is a human institution, subject to human foibles — and fragile, because it depends as much on group ethics as on individual ethics. It can break down, and I believe it is doing so now.

A community often finds it is constrained to think in a particular way because of how it is organized. An important organizational issue is: Are we recognizing and rewarding the right kind of physics, and the right kind of physicist, in order to solve the problem at hand? Its cognitive counterpart is: Are we asking the right questions?

The one thing everyone who cares about fundamental physics seems to agree on is that new ideas are needed. From the most skeptical critics to the most strenuous advocates of string theory, you hear the same thing: We are missing something big. It was the perception of the need for something new that led the organizers of the 2005 annual Strings conference to offer a session on "The Next Superstring Revolution." And although there is currently more confidence among practitioners in other fields, every physicist I know will agree that probably at least one big idea is missing.

How do we find that missing idea? Clearly, someone has to either recognize a wrong assumption we have all been making or ask a new question, so that's the sort of person we need in order to ensure the future of fundamental physics. The organizational issue is then clear: Do we have a system that allows someone capable of ferreting out that wrong assumption or asking that right question into the community of people we support and (equally important) listen to? Do we embrace the creative rebels with this rare talent, or do we exclude them?

It goes without saying that people who are good at asking genuinely novel but relevant questions are rare, and that the ability to look at the state of a technical field and see a hidden assumption or a new avenue of research is a skill quite distinct from the workaday skills that are a prerequisite for joining the physics community. It is one thing to be a craftsperson, highly skilled in the practice of one's craft. It is quite another thing to be a seer.

This distinction does not mean that the seer is not a highly trained scientist. The seer must know the subject thoroughly, be able to work with the tools of the trade, and communicate convincingly in its language. Yet the seer need not be the most technically proficient of physicists. History demonstrates that the kind of person who becomes a seer is sometimes mediocre when compared with the mathematically clever scientists who excel at problem solving. The prime example is Einstein, who apparently couldn't get a decent job as a scientist when he was young. He was slow in argument, easily confused; others were much better at mathematics. Einstein himself is said to have remarked, "It's not that I'm so smart. It's just that I stay with problems longer."[1] Niels Bohr was an even more extreme case. Mara Beller, a historian who has studied his work in detail, points out that there was not a single calculation in his research notebooks, which were all verbal argument and pictures.[2] Louis de Broglie made the astounding suggestion that if light is a particle as well as a wave, perhaps electrons and other particles also behave as waves. He proposed this in a 1924 PhD thesis that did not impress his examiners and would have failed without the endorsement of Einstein. As far as I know, he never did anything nearly as influential in physics again. There is only one person I can think of who was both a visionary and the best mathematician of his day: Isaac Newton; indeed, almost everything about Newton is singular and inexplicable.

As noted in the preceding chapter, Thomas Kuhn made a distinction between "normal science" and scientific revolutions. Normal science is based on a paradigm, which is a well-defined practice with a fixed theory and a fixed body of questions, experimental methods, and calculational techniques. A scientific revolution happens when the paradigm breaks down, which is to say, when the theory it is based on fails to predict or explain the results of the experiments.[3] I don't think science always works this way, but there are certainly normal and revolutionary periods, and science is done differently during them. The point is that different kinds of people are important in normal and revolutionary science. In the normal periods, you need only people who, regardless of their degree of imagination (which may well be high), are really good at working with the technical tools — let us call them master craftspeople. During revolutionary periods, you need seers, who can peer ahead into the darkness.

Master craftspeople and seers come to science for different reasons. Master craftspeople go into science because, for the most part, they have discovered in school that they're good at it. They are usually the best students in their math and physics classes from junior high school all the way up to graduate school, where they finally meet their peers. They have always been able to solve math problems faster and more accurately than their classmates, so problem solving is what they tend to value in other scientists.

Seers are very different. They are dreamers. They go into science because they have questions about the nature of existence that their schoolbooks don't answer. If they weren't scientists, they might be artists or writers or they might end up in divinity school. It is only to be expected that members of these two groups misunderstand and mistrust each other.

A common complaint of the seers is that the standard education in physics ignores the historical and philosophical context in which science develops. As Einstein said in a letter to a young physicist who had been thwarted in his attempts to add philosophy to his physics courses:

> I fully agree with you about the significance and educational value of
> methodology as well as history and philosophy of science. So many

people today — and even professional scientists — seem to me like someone who has seen thousands of trees but has never seen a forest. A knowledge of the historical and philosophical background gives that kind of independence from prejudices of his generation from which most scientists are suffering. This independence created by philosophical insight is — in my opinion — the mark of distinction between a mere artisan or specialist and a real seeker after truth.[4]

Of course, some people are mixtures of both. No one makes it through graduate school who is not highly competent on the technical side. But the majority of theoretical physicists I know fall into one or the other group. What about me? I think of myself as a would-be seer who fortunately was good enough at my craft to contribute occasionally to the problem solving.

When I first encountered Kuhn's categories of revolutionary and normal science as an undergraduate, I was confused, because I couldn't tell which period we were in. If I looked at the kinds of questions that remained open, we were clearly partway through a revolution. But if I looked at how the people around me worked, we were just as obviously doing normal science. There was a paradigm, which was the standard model of particle physics and the experimental practices that had confirmed it, and it was normally progressing.

Now I understand that the confusion was a clue to the crisis I have been exploring in this book. We are indeed in a revolutionary period, but we are trying to get out of it using the inadequate tools and organization of normal science.

This, then, is my basic hypothesis about the last twenty-five years of physics. There can be no doubt that we are in a revolutionary period. We are horribly stuck, and we need real seers, and badly. But it has been a long time since seers were needed. We had a few monumental visionaries at the beginning of the twentieth century: Einstein above all, but also Bohr, Schrödinger, Heisenberg, and a few others. They failed to complete the revolution they started, but they created partially successful theories — quantum mechanics and general relativity — for us to build on. The development of these theories required a lot of hard technical work, and so for several generations physics was "normal science" and was dominated by master craftspeople. Indeed, the transition from dominance by Europeans to dominance by Americans, which took place in the 1940s, was very much the triumph of master craftspeople over seers. As noted, it brought about a reversal in the style of theoretical physics, from the reflective foundational mode of Einstein and his peers to the pragmatic, aggressive mode that gave us the standard model.

When I learned physics in the 1970s, it was almost as if we were being taught to look down on people who thought about foundational problems. When we asked about the foundational issues in quantum theory, we were told that no one fully understood them but that concern with them was no longer part of science. The job was to take quantum mechanics as given and apply it to new problems. The spirit was pragmatic; "Shut up and calculate" was the mantra. People who couldn't let go of their misgivings over the meaning of quantum theory were regarded as losers who couldn't do the work.

As someone who came into physics from reading Einstein's philosophical musings, I couldn't accept that reasoning, but the lessage was clear, and I followed it as best as I could. You could make a career only by working within quantum theory as given, not by questioning it. A fortunate circumstance won me some time at the Institute for Advanced Study in Princeton, but there was no memory there of Einstein's way of doing science — just an empty bronze likeness gazing silently out over the library.

But the revolution was not finished. The standard model of particle physics was certainly the triumph of this pragmatic style of doing physics, but its triumph seems now to have also marked its limit. The standard model, and just possibly inflation, is about as far as we could go with normal science. Since then, we have been mired, because what we need is a return to a revolutionary kind of science. Once again, we need a few seers. The problem is that there are now very few around, as a result of science having been done so long in a way that rarely recognized and barely tolerated them.

Between the early twentieth century and the last quarter century, science — and the academy in general — has become much more organized and professionalized. This means that the practice of normal science has been enshrined as the single model of good science. Even if everyone can see that a revolution is necessary, the most powerful parts of our community have forgotten how to make one.

We have been trying to do so with structures and styles of research best suited to normal science. The paradoxical situation of string theory — so much promise, so little fulfillment — is exactly what you get when a lot of highly trained master craftspeople try to do the work of seers.

I'm sure that some string theorists will object to this characterization. Certainly they work on fundamental problems of physics, and all their work is aimed at discovering new laws. Why are string theorists not seers? Aren't wormholes, higher dimensions, and multiple universes imaginative ideas? Yes, of course, but this is not the point. The question is: What is the context, and what are the ideas about? Hidden dimensions and wormholes are hardly a novelty more than three quarters of a century after Kaluza and Klein. Nor does it take much foresight or courage to think about these things when hundreds of other people are thinking the same thoughts.

Another way to look at our present situation is that seers are compelled, by their desire for clarity, to grapple with the deepest problems in the foundations of physics. These include the foundations of quantum mechanics and the problems associated with space and time. Many papers and books have been written on the foundational problems of quantum mechanics during the last few decades, but, to my knowledge, not a single one is by a leading string theorist. Nor do I know of any paper by a string theorist that attempts to relate the issues faced by string theory to the older writings by physicists and philosophers on the big issues in the foundations of space, time, or quantum theory.

The leaders of the background-independent approaches to quantum gravity tend, by contrast, to be people whose scientific views were formed by lifelong reflection on the deep foundational issues. It is easy to list those whose thinking has led to papers and even books addressing foundational issues: Roger Penrose is perhaps the best known to the public, but one can name many others including John Baez, Louis Crane, Bryce DeWitt, Fay Dowker, Christopher Isham, Fotini Markopoulou, Carlo Rovelli, Rafael Sorkin, and Gerard 't Hooft.

By contrast, I can think of no mainstream string theorist who has proposed an original idea about the foundations of quantum theory or the nature of time. String theorists tend to answer this charge with a dismissive response, to the effect that these questions are all solved. Occasionally they acknowledge that the problems are serious but quickly follow this admission with the claim that it's too soon to try solving them. Often one hears that we should just continue to follow the development of string theory, because since string theory is right, it must contain the necessary solutions.

I have nothing against people who practice science as craft, whose work is based on the mastery of technique. This is what makes normal science so powerful. But it is a fantasy to imagine that foundational problems can be solved by technical problem solving within existing theories. It would be nice if this were the case — certainly, we would all have to think less, and thinking is really hard, even for those who feel compelled to do it. But deep, persistent problems are never solved by accident; they are solved only by people who are obsessed with them and set out to solve them directly. These are the seers, and this is why it is so crucial that academic science invite them in rather than exclude them.

Science has never been organized in a way that is friendly to seers; Einstein's employment situation is hardly a lone example. But a hundred years ago, the academy was much smaller and much less professional, and well-trained outsiders were common. This was a legacy of the nineteenth century, when most of the people who did science were enthusiastic amateurs, either rich enough to not need to work or convincing enough that they could find a patron.

Fine, you might say. But who are the seers? They are by definition highly independent and self-motivated individuals who are so committed to science that they will do it even if they can't make a living at it. There should be a few out there, even though our professionalized academy is unfriendly to them. Who are they and what have they managed to do to solve the big problems?

They are hiding in plain sight. They can be recognized by their rejection of assumptions that most of the rest of us believe in. Let me introduce you to a few of them.

I have a lot of trouble believing that special relativity is false; if it is, then there is a preferred state of rest and both the direction and speed of motion must be ultimately detectable. But there are a few theorists around who have no trouble with this concept. Ted Jacobson is a friend who collaborated with me on a paper on the quantum mechanics of loop quantum gravity. Together we found the first exact solutions to a key equation known as the Wheeler-DeWitt equation.[5] But when loop quantum gravity surged ahead, Jacobson grew pessimistic. He didn't think loop quantum gravity would work, and he also didn't think it went deeply enough. After mulling this over, he began to question the relativity principle itself and to believe in the possibility of a preferred state of rest. He has spent years developing this idea. In chapters 13 and 14, I noted that if special relativity is wrong, experiment may soon tell us. Jacobson and his students at the University of Maryland are among the leaders of the search for an experimental test of special relativity.

Another seer who has questioned the whole framework of relativity is the cosmologist João Magueijo (see chapter 14). He had no choice, because he had invented, and fallen in love with, an idea that seemed to contradict it — which is that the speed of light might have been much faster in the early universe. The papers he wrote about this are just barely consistent — and they certainly make no sense unless one assumes that the relativity principle needs to be thrown out, or at least modified.

Then there are the wild guys of solid-state physics — accomplished physicists who made great careers explaining real things about the behavior of real stuff. I am referring to Robert Laughlin, who was awarded a Nobel Prize in 1998 for his contributions to "the discovery of a new form of quantum fluid with fractionally charged excitations," Grigori Volovik, of the Landau Institute for Theoretical Physics, in Moscow, who explained the behavior of certain species of very cold liquid helium, and Xiao-Gang Wen. These men are master craftsmen and seers both. Having done perhaps the best and most consequential normal science of the last few decades, they decided to try their hands at the deep problems of quantum gravity, and they started with the idea that the relativity principle is false, that it is just an approximate, emergent phenomenon. The particle physicist James Bjorken is another such seer/craftsperson. That we know that protons and neutrons contain quarks is due in large part to his insights.

One of the great seers is Holger Bech Nielsen, of the Niels Bohr Institute. He was an inventor of string theory, and he has many other key discoveries to his credit. But for many years he has been isolated from the mainstream for advocating what he calls *random dynamics*. He believes that the most useful assumption we can make about the fundamental laws is that they are random. Everything we think of as intrinsically true, such as relativity and the principles of quantum mechanics, he thinks are just accidental facts that are emergent from a fundamental theory so beyond our imagining that we might as well assume that its laws are random. His models are the laws of thermodynamics, which used to be based on principles but now are understood as the most likely way that large numbers of atoms in random motion will behave. This may not be right, but Nielsen has come remarkably far in his antiunification program.

There is a very short list of string theorists who have made lasting contributions to science on the level of those devised by these gentlemen. So how do the string theorists — or for that matter, the loop theorists — respond to the insistent warnings of these accomplished physicists that perhaps we are all making a wrong assumption? We ignore them. Yes, really, flat out. To tell the truth, we laugh at them behind their backs, and sometimes as soon as they have left the room. Having done Nobel Prize–level physics — or even having won the prize itself — apparently doesn't protect you when you question universally held assumptions such as the special and general theories of relativity. I was shocked when Laughlin told me that he was under pressure from his department and funding agency to keep doing normal science in the field he had been working in, rather than spending time on his new ideas about space, time, and gravity. If such a person, after all his accomplishments, including the Nobel, cannot be trusted to chase down his deepest ideas, what exactly is the meaning of academic freedom?

Fortunately for physics, we will soon know whether special relativity is true or not. Most of my friends expect that experimental observations will prove these great men fools. I hope the iconoclasts are wrong and that special relativity passes the test. But I cannot rid myself of the fear that perhaps we are the ones who are wrong and they are right.

So much for questioning relativity. What if quantum theory is wrong? This is the soft underbelly of the whole project of quantum gravity. If quantum theory is wrong, then trying to combine it with gravity will have been a huge waste of time. Does anyone think this is the case?

Yes, and one is Gerard 't Hooft. As a graduate student at the University of Utrecht, 't Hooft proved, with an older collaborator, that quantum Yang-Mills theories were sensible, a discovery that made the whole standard model possible, and he has a well-deserved Nobel Prize for these efforts. That's only one of his many fundamental discoveries about the standard model. But for the last decade he has been one of the boldest thinkers on foundational issues. His main idea is called the *holographic principle*. As he formulates it, there is no space. Everything that happens in a region we are used to thinking of as space can be represented as taking place on a surface surrounding that space. Furthermore, the description of the world that exists on that boundary is not quantum theory but a deterministic theory he believes will replace it.

Just before 't Hooft formulated his principle, a similar idea was proposed by Louis Crane in the context of background-independent approaches to quantum gravity. He proposed that the right way to apply quantum theory to the universe is not to try to put the whole universe into one quantum system. This had been tried by Stephen Hawking, James Hartle, and others and found to face severe problems. Crane suggested instead that quantum mechanics is not a static description of a system but a record of information that one subsystem of the universe can have about another by virtue of their interaction. He then suggested that there is a quantum-mechanical description connected with every way of dividing the universe into two parts. The quantum states live not in one part or the other but on the boundary between them.[6]

Crane's radical suggestion has since grown into a class of approaches to quantum theory that are called *relational quantum theories*, because they are based on the idea that quantum mechanics is a description of relationships between subsystems of the universe. This idea was developed by Carlo Rovelli, who showed it to be perfectly consistent with how we usually do quantum theory. In the context of quantum gravity, it resulted in a new approach to quantum cosmology, made by Fotini Markopoulou and her collaborators. Markopoulou emphasized that describing the exchange of information between different subsystems is the same as describing the causal structure that limits which systems can influence each other. She thus found that a universe can be described as a quantum computer, with a dynamically generated logic.[7] The idea that the universe is a kind of quantum computer has also been promoted by Seth Lloyd of MIT, one of the visionaries of the field of quantum computation.[8] From the two sides of their respective disciplines, Markopoulou and Lloyd have been leading a movement that uses ideas from quantum information theory to reconceptualize the universe, leading to the understanding of how elementary particles can emerge from quantum spacetime.

Gerard 't Hooft's idea of a world represented on its boundary should remind you of the Maldacena conjecture. Indeed, 't Hooft's ideas were in part an inspiration for Juan Maldacena, and some think the holographic principle will turn out to be one of the basic principles of string theory. This alone could have easily made him one of the leaders of the string theory community, had he been interested in such a role. But in the 1980s, 't Hooft began to go his own way. He did this while he was in his prime and at a point when no one was technically stronger. Still, the minute he deviated from the mainstream, he was laughed at by his fellow particle physicists. He didn't seem to care, or even to notice, but I'm sure it stung. Nevertheless, he doubted almost everything and forged his own path in fundamental physics. His core belief, developed over decades, is that quantum physics is wrong.

There is no more earnest or sincere person than 't Hooft. One thing we in the field of quantum gravity love about him is that he is so often there. He comes to many of our meetings, and there you never see him in the halls, politicking with the other prominent attendees. Instead, he comes to every session, something only the young students do. He arrives first thing each morning, impeccably dressed in a three-piece suit (the rest of us are generally in jeans and T-shirts), and he sits in the front row all day and listens to the talks by every single student and postdoc. He doesn't always comment, and he may even doze off for a minute or two, but the respect he shows by being there for each of his colleagues is impressive. When it's his turn to speak, he stands up and unpretentiously presents his ideas and results. He knows that his is a lonely road, and I would not be surprised if he resents it. How does a person give up the mantle of leadership, so richly deserved, just because he can't make sense of quantum mechanics? Imagine what that says about someone's character.

Then there's Roger Penrose. Simply put, there is no one who has contributed more to our understanding and use of the general theory of relativity, save Einstein himself, than Roger Penrose. He is one of the four or five most talented and deeply original thinkers I have met in any field. He has done great mathematics and great physics. Like 't Hooft, much of his work in the last two decades is motivated by his conviction that quantum mechanics is wrong. And like 't Hooft, he has a vision of what should replace it.

Penrose has been arguing for years that the incorporation of gravity into quantum theory makes that theory nonlinear. This leads to a resolution of the measurement problem, in that quantum-gravitational effects cause the quantum state to collapse dynamically. Penrose's proposals are well described in his books, although they have not so far been implemented in a detailed theory. Nevertheless, he and others have been able to use them to make predictions for doable experiments, some of which are presently under development.

A few of us take Penrose's arguments seriously; an even smaller number are convinced of their validity. But most string theorists — and certainly all mainstream string theorists — show no signs of listening at all. If even the most honored visionaries are not taken seriously once they begin to question basic assumptions, you can imagine how well people fare who are seers but not lucky enough to have made substantial contributions first.[9]

If several of the best living theoretical physicists feel compelled to question the basic assumptions of relativity and quantum theory, there must be others who come to this position from the beginning. There are indeed people who, early in their studies, began to think quantum theory must be wrong. They learned it, and they can carry out its arguments and calculations as well as anyone. But they don't believe in it. What happens to them?

There are roughly two kinds of such people: the sincere ones and the insincere ones. I am one of those who never found a way to believe in quantum mechanics, but I am one of the insincere ones. That is, I understood early in my education that I could not have a decent career as an academic theoretical physicist if I focused on trying to make sense of quantum mechanics. So I decided to do something that the mainstream would understand and appreciate well enough so that I could pursue a normal career.

Luckily I found a way to investigate my doubts about the foundations by working on something as mainstream as quantum gravity. Since I didn't believe in quantum mechanics in the first place, I was pretty sure that this effort had to fail, but I hoped that understanding the failure would provide clues as to what might replace quantum theory. A few years earlier, I would have had as little luck with a career based on quantum gravity as with one based on worrying about quantum theory being wrong. However, an easy opportunity opened up while I was a graduate student, which was to attack the problem of quantum gravity using recent methods developed to study the standard model. So I could pretend to be a normal-science kind of physicist and train as a particle physicist. I then took what I learned and applied it to quantum gravity. Since I was among the first to try this approach, and since I used tools that the leaders of the mainstream understood, this made a decent, if not stellar, career possible.

But I could never completely suppress the instinct to probe the foundations of my subject. I wrote a paper in 1982 titled "On the Relationship Between Quantum and Thermal Fluctuations" that, when I look at it now, seems unbelievable to me in its boldness.[10] I asked a new question about how space, time, and the quantum fit together — a question that opened up a whole new way to tackle the problem. Even now, after having written many influential papers, I think this one was my best. Every once in a while, I meet a student who is reading back into the foundations of the subject, or some loner who has been on the outskirts for decades, and they say, "Oh, you're that Smolin! I never made the connection. I thought he must have died, or left physics." Now at last, along with my colleagues at Perimeter, I am finally returning to work on the foundations of quantum mechanics.

What of the sincere people, who didn't believe basic assumptions like relativity and quantum theory and didn't have a malleable enough character to suppress their inclinations? They are a special breed, and they each have a story to tell.

Julian Barbour is known to many who follow science as the author of *The End of Time*, in which he argues that time is an illusion.[11] He is an unusual physicist, who, since receiving his doctorate in 1968 from the University of Cologne, has never held an academic job. But he has been highly influential among the small group of people who think seriously about quantum gravity, for it was he who taught us what it means to make a background-independent theory.

As Barbour tells it, on a climbing trip during graduate school, he was seized by a vision that time might be an illusion. This led him to investigate the roots of our understanding of time, contained in the general theory of relativity. He realized that he could not make a conventional academic career worrying about the nature of time. He also realized that if he was going to work on that problem, he would have to concentrate on it fully, without being distracted by the pressures of a normal career in physics. So he bought an old farmhouse in a little village half an hour from Oxford, brought his new wife there, and settled down to think about time. It was ten years or so before he had something to report back to his colleagues. During this period, he and his wife had four children, and he worked part-time as a translator to support them. The translating took him no more than twenty hours a week, leaving him as much time for thinking as most academic scientists have after the responsibilities of teaching and administration are taken into account.

To get a bearing on the meaning of time in general relativity, Barbour read deeply into the subject, working his way back through the history of physics and philosophy. He finally was able to invent a new kind of theory, in which space and time are nothing but a system of relationships. His papers on this subject slowly began to be noticed, and eventually he became an honored member of the quantum-gravity community. His reinterpretation of Einstein's general theory of relativity as a relational theory is now the way we in the field understand it.

This is not nearly all that Barbour has done, but it's enough to show how the career of a successful seer differs from that of a conventional academic scientist. Such a person does not follow fashion — in fact, probably does not even follow a field well enough to know what the fashion is. People like this are driven by nothing except a conviction, gained early, that everyone else is missing something crucial. Their approach is more scholarly, in that to think clearly they have to read through the whole history of the question that obsesses them. Their work is intensely focused, yet it takes them a long time to get somewhere. In furtherance of an academic career there is no output whatsoever. Julian Barbour, when he was ready, changed science more than most academic scientists have, but at an age when most academic physicists are up for tenure, he had absolutely nothing to show for his work.

Barbour's career resembles that of other seers, like Charles Darwin, who also retreated to the English countryside to find the room to think through an idea that obsessed him. Einstein spent ten years thinking about the ideas that became special relativity, and then spent the next ten inventing general relativity. Time and the freedom to think, then, are all that a seer needs to find that unexamined assumption. The rest they do themselves.

Another such person is David Finkelstein, professor emeritus of the Georgia Institute of Technology, who has spent his whole life looking for the logic of nature. He does physics differently from anyone else. His life's work has been a quest to understand, as he put it when we first met, "how God might have thought the world into existence." He never did anything but this, and each time we meet he has a new insight about it. Along the way, there have been a few spin-offs. He was the first person to understand what the event horizon of a black hole is.[12] He was the first to discover important features of solid-state physics called topological conservation laws, and he was also the first to study a variety of mathematical structures — quantum groups, for example. His life serves as a lesson for the range of contributions a seer can make while on his own road to truth. Whereas Finkelstein did have an academic career, could someone like him — someone who listens only to an inner voice and ignores almost everything else — get a professorship these days at a major university? Dream on.

Here is another story, this one more like Barbour's. Antony Valentini started with an undergraduate degree from Cambridge, as did Barbour. Then he wandered Europe for a few years, eventually settling in Trieste to study with Dennis Sciama, who at Cambridge had been the teacher of Stephen Hawking, Roger Penrose, Martin Rees,

George Ellis, and several other great relativists and cosmologists. Late in his career, Sciama had moved to Trieste and founded an astrophysics group at a new Italian institute called SISSA (for Scuola Internationale Superiore di Studi Avanzati). Valentini was one of the last of Sciama's students, and he didn't work on astrophysics; instead, he pursued work in quantum theory, based on a gut feeling that it made no sense. He studied an old idea, first developed by Louis de Broglie in the 1920s, called a *hidden-variables theory*, according to which there is a single reality hidden behind the equations of quantum theory. The idea of hidden variables was suppressed for decades — despite support by Einstein, Schrödinger, and others — partly because of a false proof published by John von Neumann in 1932 that such theories could not exist. The mistake was finally uncovered by the quantum theorist David Bohm in the early 1950s, who then revived de Broglie's theory. Valentini made a new and very important modification of the hidden-variables theory, the first improvement in that theory in decades. Most of his papers on this were rejected by the physics journals, but their contents are now widely accepted among specialists who work on the foundations of quantum mechanics.

Sciama did what he could to encourage and help Valentini, but there were no academic positions available either in Italy or the English-speaking world for someone whose work focused on foundational problems. Sciama did suggest to Valentini that if he couldn't publish his growing body of results in journals, he should write a book about them. Without a position, Valentini moved to Rome, where he eventually secured a postdoc at the University of Rome. When that ran out, he stayed in Rome for six more years, in love with the city and one of its inhabitants, supporting himself by tutoring and meanwhile developing his theory and putting the results into his book.[13]

While many leading physicists admit private misgivings about quantum mechanics, their public stance is that its problems were settled back in the 1920s. A scholarly account of the later work on its foundations does not exist, but I know that since at least the 1950s, the leading journals have only very selectively published papers on this subject, while several journals have excluded such papers by stated policy. The granting agencies and major government foundations have typically not supported this work,[14] and university departments have tended not to hire people who are doing it.

This general recalcitrance is partly a result of the move from revolutionary science to normal science in the 1940s. As in a political revolution, the rebellion had to be suppressed if the revolution was to consolidate its gains. In the early years, there had been several competing views and ideologies about the interpretation of quantum theory. By the 1940s, one had triumphed. In deference to the leadership of Niels Bohr, it was called the Copenhagen interpretation. Bohr and his followers had a stake in cutting off debate, and I would not be surprised to learn that they used the levers of academic politics to do so; given their involvement in the invention of nuclear weapons, they were certainly well placed to succeed. But even those who didn't care about ideology and wanted only to get on with doing normal science had a motive for stifling debate on the subject. Quantum theory was, on the experimental and practical side, a great success, and those forging ahead with it did not want to be bothered by the nagging doubts of those who continued to worry that there were deep problems with how the theory was formulated and interpreted. It was time to move on.

Those doubters who persevered had few choices. Some restyled themselves as philosophers and published long scholarly arguments in philosophy journals. They created a little subculture that at least kept the debate alive. A few who had mathematical talent got jobs in mathematics departments, where they published formal, rigorous work on alternatives to the consensus formulation of quantum mechanics. Others — some of the best people in the field — found professorships at small colleges, where you were not required to get research grants. A few others made physics careers based on work in other fields and from time to time worked on quantum mechanics as a kind of hobby.

One of these "hobbyists" was John Stewart Bell, who discovered a key theorem about hidden-variable theories in the early 1960s. He built his career on good work in particle physics, but now, some years after his death, it is clear that his most important contribution was his work on quantum theory. Bell is sometimes quoted as having said that one should do normal science and spend only 10 percent of one's time worrying about the quantum theory. When this dictum comes up, my Perimeter colleague Lucien Hardy likes to speculate on how much more Bell might have contributed to science if he had spent more time in the area where he made the biggest impact — except then he would likely have had no job at all.

It is not surprising that through this period very little progress was made on the foundations of quantum mechanics. How could it have been otherwise? Of course, this was often reason enough not to hire, fund, or publish the few people who did make progress.

Now we know how wrong the skeptics were. About twenty years ago, Richard Feynman and a few others realized that you might be able to make a new kind of computer using quantum phenomena in an essential way. The suggestion was largely unexplored until a more detailed proposal for a quantum computer was made in 1985 by David Deutsch, now at the Centre for Quantum Computation at Oxford.[15] There is no more foundational thinker than Deutsch; he was motivated to invent quantum computers by his disquiet with foundational problems in both mathematics and quantum theory. Just how original and clear a thinker he is can be seen in his provocative book *The Fabric of Reality*,[16] in which he elaborates on his many-worlds theories. I disagree with much of what he writes, but I loved it.

In 1994, Peter Shor of MIT, who was then a computer scientist at Bell Laboratories, found a remarkable result, which is that a large enough quantum computer would be able to break any code in existence.[17] Since then, money has flooded into the field of quantum computation, as governments do not want to be the last to have their codes broken. This money has supported a new generation of young, very smart scientists — physicists, computer scientists, and mathematicians. They have created a new field, a blending of physics and computer science, a significant part of which involves a reexamination of the foundations of quantum mechanics. All of a sudden, quantum computing is hot, with lots of new ideas and results. Some of these results address the concerns about the foundations, and many could have been discovered anytime since the 1930s. Here is a clear example of how the suppression of a field by academic politics can hold up progress for decades.

In 1999, after seven years of isolation in Rome, Antony Valentini moved back to his parents' house in London. His family had emigrated from a small village in Abbruzzo; they owned a little store, and they were willing to support him in his work for as long as it took. I met him there that year, when I was a visiting professor at Imperial College, and after discussions with Christopher Isham, head of the theory group there, we decided to offer him a postdoc and bring him back into science. We were able to do this because I had some unexpected and generous support from a donor who happened to care about the foundations of quantum mechanics. I felt that supporting one of the few people who had proved he could contribute new and important results to this field was putting that money to good use. Had I been supported only by funds from the National Science Foundation, I would not have been able to do this. As generous as the NSF has been to me for my work on quantum gravity, sharing the grant with a postdoc working on foundations of quantum theory could have hurt the chances for future funding.

Now Valentini has joined us at Perimeter. He is still working on his hidden-variables book, but in the meantime he has become a leading figure in the field of foundations of quantum theory, an invited speaker at many conferences on the subject. He now publishes regularly, and his most recent work concerns a bold new proposal to test quantum mechanics by observing X rays that originate near black holes.[18] Like Julian Barbour, his years of isolation allowed him to engage in scholarly self-education, and there isn't a more insightful or knowledgeable critic in the whole field of quantum theory.

Keep in mind why Barbour and Valentini could not have accomplished anything had they tried to have an ordinary academic career. During the stage when one is normally an assistant or associate professor, working hard to be published and renowned enough to win the invitations and research grants necessary to get tenure, they were publishing nothing. But they were accomplishing a great deal. They were thinking, and in a deeper, more focused way than an assistant professor can, about a single recalcitrant foundational issue. When they emerged, after roughly a decade, each had a considered, original, and mature viewpoint that led to their quickly becoming influential. The authority gained from having gone through these years of concentrated study and thought and come out of it with something new and important made them essential to people who cared about these issues.

For seers, the need to be alone for an extended period at the beginning of a career, and often in later periods, is essential. Alexander Grothendieck is said by some to be the most powerful and visionary mathematician now alive. He has had a most unconventional career. Some of his major contributions, which were seminal, were not published, but mailed — in the form of letters hundreds of pages long sent to friends and then passed hand to hand among small circles of people who could read them. His parents were refugees from political oppression and war; he grew up in refugee camps after the Second World War. He appeared in the mathematical world of Paris as if from nowhere. After a brief but extraordinarily influential career, he largely withdrew from scientific life in the 1970s, at least partly because he objected to military funding for mathematics. He disappeared altogether in 1991, and although rumor has it that he is living as a hermit in the Pyrenees, his whereabouts are still uncertain. Clearly, he is an extreme case. But you have to see the look of admiration, wonder, and perhaps even a little fear on the faces of some very good mathematicians whenever his name comes up. Here is how he describes some of his experiences:

In those critical years I learned how to be alone. [But even] this formulation doesn't really capture my meaning. I didn't, in any literal sense learn to be alone, for the simple reason that this knowledge had never been unlearned during my childhood. It is a basic capacity in all of us from the day of our birth. However these three years of work in isolation [1945–1948], when I was thrown onto my own resources, following guidelines which I myself had spontaneously invented, instilled in me a strong degree of confidence, unassuming yet enduring, in my ability to do mathematics, which owes nothing to any consensus or to the fashions which pass as law. . . . By this I mean to say: to reach out in my own way to the things I wished to learn, rather than relying on the notions of the consensus, overt or tacit, coming from a more or less extended clan of which I found myself a member, or which for any other reason laid claim to be taken as an authority. This silent consensus had informed me, both at the lycée and at the university, that one shouldn't bother worrying about what was really meant when using a term like "volume," which was "obviously self-evident," "generally known," "unproblematic," etc. . . . It is in this gesture of "going beyond," to be something in oneself rather than the pawn of a consensus, the refusal to stay within a rigid circle that others have drawn around one — it is in this solitary act that one finds true creativity. All others things follow as a matter of course.

Since then I've had the chance, in the world of mathematics that bid me welcome, to meet quite a number of people, both among my "elders" and among young people in my general age group, who were much more brilliant, much more "gifted" than I was. I admired the facility with which they picked up, as if at play, new ideas, juggling them as if familiar with them from the cradle — while for myself I felt clumsy, even oafish, wandering painfully up an arduous track, like a dumb ox faced with an amorphous mountain of things that I had to learn (so I was assured), things I felt incapable of understanding the essentials or following through to the end. Indeed, there was little about me that identified the kind of bright student who wins at prestigious competitions or assimilates, almost by sleight of hand, the most forbidding subjects.

In fact, most of these comrades who I gauged to be more brilliant than I have gone on to become distinguished mathematicians. Still, from the perspective of thirty or thirty-five years, I can state that their imprint upon the mathematics of our time has not been very profound. They've all done things, often beautiful things, in a context that was already set out before them, which they had no inclination to disturb. Without being aware of it, they've remained prisoners of those invisible and despotic circles which delimit the universe of a certain milieu in a given era. To have broken these bounds they would have had to rediscover in themselves that capability which was their birthright, as it was mine: the capacity to be alone.[19]

It is a cliché to ask whether a young Einstein would now be hired by a university. The answer is obviously no; he wasn't even hired then. Now we are much more professionalized, and hiring is based on stringent competition among people highly trained in narrow technical skills. But some of the others I've mentioned could not be hired either. If we have the contributions of these people, it is because of their generosity — or maybe their stubbornness — in continuing to work without the support the academic world normally gives to scientists.

At first this would seem easy to correct. There are not very many such people, and they are not hard to recognize. Few scientists think about foundational problems, and even fewer have ideas about them. My friend Stuart Kauffman, the director of the Institute for Biocomplexity and Informatics at the University of Calgary, once told me that it's not hard to pick out the people with daring ideas — they have almost always had at least a few such ideas already. If they haven't had any by the end of graduate school or a few years later, they probably never will. So how do you distinguish between the seers who have good ideas and others who try but just don't? This is easy too. Just ask older seers. At Perimeter, we have no problem picking out the few young ones who are worth watching.

But once these people are identified, they must be treated differently from those doing normal science. Most of them are uninterested in who is cleverer, or who is quicker at solving problems presented by mainstream normal science. And if they tried to compete, given how strict the competition is, they would fail. If they are competing with anyone, it is with the last generation of revolutionaries, who speak to them from the old books and papers that no one else ever reads. There is little external that drives them; they are focused on inconsistencies and issues in science that most scientists are willing to ignore. If you wait five years or even ten, they are not going to look good according to the usual criteria. You can't panic, but you do have to leave them alone. Eventually, like Barbour and Valentini, they will emerge with something that has been worth the wait.

As there are not many such people, it should not be hard to make room for them in the academy. Indeed, you would think that many institutes, colleges, and universities would be happy to have such people. Because they think clearly about the foundations of their subjects, they are often good, even charismatic teachers. Nothing inspires students like a seer on fire. Because they are not competitive, they are good advisors and mentors. After all, isn't the main business of colleges and universities to teach?

Of course, there is a real risk. Some of them will not discover anything. I am talking in terms of a real lifetime contribution to science. But then most academic scientists, though they succeed in career terms — get grants, publish a lot of papers, go to a lot of conferences, and so on — contribute only incrementally to science. At least half our colleagues in theoretical physics fail to make a unique or genuinely lasting contribution. There is a difference between a good career and an essential career. Had they done something else with their lives, science would have gone on much the same. So it's a risk either way.

The nature and costs of different kinds of risk are issues that businesspeople understand better than academic administrators. It is much easier to have a useful and honest conversation about this with a businessperson than with an academic. I once asked a successful venture capitalist how his company decided how much risk to take on. He said that if more than 10 percent of the companies he funded made money, he knew he was not taking on enough risk. What these people understand, and live with, is that you get overall a maximal return, which maps to a maximal rate of technological progress, when 90 percent of new companies fail.

I wish I could have an honest conversation about risk with the National Science Foundation. Because I'm sure that 90 percent of the grants they give in my field fail, when measured against the real standard: Do those grants lead to progress in science that would not have occurred if the person funded did not work in the field?

As every good businessperson knows, there is a difference between low-risk/low-payoff and high-risk/high-payoff strategies, starting with the fact that they are designed with different goals in mind. When you want to run an airline or a bus system or make soap, you want the first. When you want to develop new technologies, you cannot succeed without the second.

What I wouldn't give to get university administrators to think in these terms. They set up the criteria for hiring, promotion, and tenure as if there were only normal scientists. Nothing should be simpler than just changing the criteria a bit to recognize that there are different kinds of scientists, with different kinds of talents. Do you want a revolution in science? Do what businesspeople do when they want a technological revolution: Just change the rules a bit. Let in a few revolutionaries. Make the hierarchy a bit flatter, to give the young people more scope and freedom. Create some opportunities for high-risk/high-payoff people, so as to balance the huge investment you made in low-risk, incremental science. The technology companies and investment banks use this strategy. Why not try it in academia? The payoff could be discovering how the universe works.

19

How Science Really Works

THE IDEA OF CHANGING the way science is done in universities will no doubt appeal to some, while horrifying others. But it's probably in no danger of happening. To explain why, we need to inspect the dark underbelly of academic life. Because, as the sociologists tell us, it is not just about wisdom, it is about power: who has it, and how it is used.

There are certain features of research universities that discourage change. The first is peer review, the system in which decisions about scientists are made by other scientists. Just like tenure, peer review has benefits that explain why it's universally believed to be essential for the practice of good science. But there are costs, and we need to be aware of them.

I'm sure the average person has no idea how much time academics spend making decisions about hiring other academics. I spend roughly five hours a week in committees discussing other peoples' careers or writing letters to be read by such committees. I have been doing this for some time. It is a major part of the job of a professor, and many professors I know spend more time on it than I do. One thing is sure: Unless you embarrass yourself by a visible show of irresponsibility or prove yourself too unpredictable or too untrustworthy, the longer you are a scientist, the more time you will spend meddling in the careers of other scientists. It's not just that you will have more and more students, postdocs, and collaborators who need letters written for them; you are also involved in the hiring decisions of other universities and institutes.

Has anyone in administration ever studied how much this system costs us? Is it really necessary? Could we spend less time on this and have more time for science and teaching? I have had only a small taste of the system, and it is daunting. No department or institute with aspirations can hire without consulting a network of visiting and advisory committees staffed by influential older scientists from other institutions. There are also panels set up by funding agencies in the United States, Canada, Europe, and around the globe. Then there are all the informal contacts, telephone calls, and conversations in which you are asked to frankly and confidentially evaluate lists of candidates. After a certain point, a successful scientist could easily spend all of his or her time on the politics of who gets hired where.

This is called peer review. It's a funny name, because it differs markedly from the notion of a jury of one's peers, which suggests that you are being judged by people just like yourself, who are presumably fair and objective. There are real penalties — prison — for jurors who conceal a bias.

In the academic world, with few exceptions, the people who evaluate you are older than you are, and more powerful. This is true all the way up the ladder, from your first college course to the applications you make for grants when you're a professor. I do not want to disparage the hard work done by so many in the service of peer review. Most do it sincerely. But there are big problems with it, and they are relevant to the state of physics today.

An unintended by-product of peer review is that it can easily become a mechanism for older scientists to enforce direction on younger scientists. This is so obvious that I'm surprised at how rarely it is discussed. The system is set up so that we older scientists can reward those we judge worthy with good careers and punish those we judge unworthy with banishment from the community of science. This might be fine if there were clear standards and a clear methodology to ensure our objectivity, but, at least in the part of the academy where I work, there is neither.

As we have discussed in detail, different kinds of scientists contribute to theoretical physics, and they all have different strengths and weaknesses. There is, however, little acknowledgment of this; instead, we speak simply about who is "good" and who is "not good" — that is, peer review is based on the simplistic and obviously faulty assumption that scientists can be ranked as on a ladder.

When I read my first batch of letters of recommendation as a new assistant professor at Yale in 1984, I couldn't believe it. While a good letter might convey a lot of information and make some attempt at nuance, a great deal of attention was given to the final paragraph, which usually offered a comparative ranking of the candidate: "X is better than A, B, and C, but not as good as E, F, G, and H." I've read thousands of letters of recommendation by now, and at least half contain some version of that sentence. In the early years, there were a few instances when I was person A, B, or C. I recall agreeing that candidate X was indeed better than I was — and in fact some of those Xs have gone on to do very well. But I would not be surprised if research showed that these rankings are, on average, poor predictors of genuine success in science. If we were really concerned about making good hires, we would carry out such research. Certainly, there are not a few cases in which the top-ranked postdoc or assistant professor does not end up doing much and doesn't get tenure.

What makes this practice even more problematic is that there are no sanctions for bias. A professor will shamelessly write letters slanted toward his or her own students, or for people who are following his or her particular research program, or even for people of his or her own nationality. We may notice (and laugh about) the really blatant instances, but no one thinks it unusual. It's just part of the system.

Here's a basic rule for predicting the kind of junior scientist that senior scientists will recommend: Does the junior scientist remind them of themselves? If you see a younger version of yourself in X, then X must be really good. I know I am guilty of this, and I say so frankly. If you want to hire more people like me, I am great at picking them out. If you want to make fine distinctions among people very different from me, who are good at things I am not so good at or don't value, don't trust my judgments.[1]

Even for those of us who strive to be fair, there is no guidance or training on how to be objective. I have never been given any advice about how to write or interpret letters of recommendation, nor have I ever seen guidelines on how to recognize signs of prejudice or stereotyping in your own or others' views. I have served on many committees for hiring, tenure, and promotion, but I have never been instructed, as jurors are instructed, on how to best weigh the evidence.

Once at a dinner party I asked people in other lines of work if they were trained in such matters. Everyone who was not an academic but who had responsibility for hiring or supervising other people had been given several days of training in recognizing and combating signs of unfairness or prejudice, in discounting the effects of hierarchy, and in encouraging diversity and independence of thought. They knew all about "heeding all the voices in your organization" and getting "a 360-degree look at job candidates" by seeking evaluations from people the candidates had supervised as well as from people who had supervised the candidates. If lawyers, bankers, television producers, and newspaper editors are assumed to need guidance in how to make wise and fair personnel decisions, why do we academic scientists assume we can do it automatically?

It is even worse than that. Behind the formal letters of recommendation is a network of confidential, informal conversations with experts: "What do you think of so and so? Whom do you think we should hire?"

These conversations are frank. The gloves are off. And this is not all bad. Many people try to rise to the challenge and be helpful, but the average level of objectivity is shockingly low. And here, especially, there is no price to pay for gaming the system in favor of your friends and the students of your friends. It is common for established experts to push their own students and postdocs, praising them unreasonably over others, especially the students of rivals.

Even in these frank exchanges, you seldom hear really negative comments. When people have nothing good to report, they will often just say, "Let's move on. I'd rather not comment" or something mild like "I'm not excited." But there are times when the mere mention of a name invokes an "Absolutely not!" or "Don't go there" or "Are you kidding?" or the definitive "Over my dead body!" In my experience, in every such instance the candidate fell into one and often two of the following three categories: They were

(1) female, (2) not white, and/or (3) someone inventing his or her own research program rather than following the mainstream. There are of course women and nonwhites who elicit no objections. But, again in my experience, these are cases where the candidate hews tightly to an established research program.

There is heated debate among physicists over why there are not more women or blacks in physics, compared with other fields just as challenging, such as mathematics or astronomy. I believe the answer is simple: blatant prejudice. Anyone who has served, as I have, on decades of hiring committees and hasn't seen naked prejudice in action is either blind to it or dishonest. There are rules and ethics of confidentiality that prevent me from giving examples, but there are several detailed studies that tell the story.[2]

Perhaps it's to be expected that prejudice is fierce in this field. How many leading theoretical physicists were once insecure, small, pimply boys who got their revenge besting the jocks (who got the girls) in the one place they could — math class? I was one of these, at least until I figured out what the jocks knew — that it is all about confidence. But I still recall feeling smug about my abilities in algebra, and I can report that, at least for me, the identification of math skills with maleness runs very deep. But then, why do women have less difficulty getting hired as pure mathematicians than as physicists? Because it is clearer in mathematics when you have done something good. A theorem is either proved or not proved, while the judgments that go into ranking theoretical physicists are much more diffuse, which gives more room for bias. It is not always easy, for example, to distinguish a good theorist from one who is just assertive. Note that whereas there have always been talented women musicians, the number of women hired by orchestras rose significantly when candidates began auditioning behind a screen.

This is why there is affirmative action. In all my experience, I have never seen a woman or an African American hired through an affirmative-action program who didn't strongly deserve it — that is, who wasn't already arguably the best applicant. When hiring committees are no longer composed only of white men and we stop hearing expressions of open prejudice, then we can relax affirmative action. As it stands, people who are different — who, for one reason or another, make powerful older male physicists uncomfortable —

are not hired. There is affirmative action for people who are visibly different, like women and blacks. But what about people who just *think* differently — who reject mainstream approaches in favor of their own ideas? Should there be affirmative action for them, too?

Many of us participate in peer review with the best intentions of choosing ethically and objectively. And when all else is equal, the more deserving candidate is chosen. That is, when you get down to comparing white men of about the same age and background, who are all pursuing the same research program, the system will generally pick out the one who is cleverest and works the hardest. But the problem is that you have to do a lot of winnowing before you reach the point where everything is equal. Up until that point, the process is political. It is the primary mechanism by which older and more powerful scientists exert power over younger scientists.

This makes for a process of enforced consensus, in which older scientists ensure that younger scientists follow their directions. There are some simple ways in which this power is exerted. For example, a candidate for a faculty position needs letters from a great many people, all more powerful than the candidate. One not-so-positive letter will normally kill an appointment. When I first encountered this plethora of recommendation letters, I was puzzled. Certainly you could get a good picture of a candidate from three or four letters. Why ten or fifteen, the number often required by highly ranked universities?

One reason is that the goal is not only to hire good scientists. Hiring committees, chairs, and deans often have another goal in mind, which is to raise (or in fortunate cases, preserve) the status of the department. By this I mean something more measurable than a young scientist's promise, for measures of status are given by numerical rankings. These are made by external evaluators, who combine their impressions with numbers like total grant funding and numbers of citations. Department chairs and deans have to be concerned with this, because such matters have brute financial repercussions relevant for their own careers as administrators. It is, first of all, important to hire people who are likely to win generous grant support. This immediately favors members of large established research programs over initiators of new programs. By asking for many letters, you can measure how well the potential hires are already perceived by senior scientists who matter. The goal, then, is not to hire the scientist most likely to do good science but the scientist whose acquisition will optimize the status of the department in the short term. This is why hiring committees don't agonize over long-term issues, like which candidate is most likely to have original ideas that will matter in twenty years. Instead, they want to know that ten to fifteen senior scientists think the candidate is a high-status member of their community.

But to elicit such a large number of positive letters, you have to be part of a large research program. If you are in a small program with fewer than ten senior people in a position to judge you, you may well be forced to ask for evaluations from people who disagree with what you do or whose programs are in competition with yours. So there is safety only in numbers. No wonder the big research programs dominate!

There is no doubt that this system has benefited string theory and made it more difficult for people who pursue alternative research programs. As a recent *New York Times* article noted, "[S]cientists have yet to develop more than fragments of what they presume will ultimately be a complete theory. Nevertheless, string theorists are already collecting the spoils that ordinarily go to the experimental victors, including federal grants, prestigious awards, and tenured faculty positions." David Gross, now the director of the Kavli Institute for Theoretical Physics at UC Santa Barbara, is quoted in the same article as saying, "Nowadays, if you're a hotshot young string theorist you've got it made."[3]

My point here is not to criticize string theory; the string theorists are just behaving in ways that members of any dominant research program would. The problem is that we have a system of decision making in academia that is far too vulnerable to takeover by an aggressively promoted research program regardless of the results. The same system once worked against string theorists. As journalist Gary Taubes notes,

> On August 4, 1985, I sat in the cantina at CERN drinking beer with Alvaro de Rujula. . . . De Rujula predicted that 90 percent of the theorists would work on superstrings and the connection with supersymmetry, because it was fashionable. When he intimated that this was not a healthy state, I asked him what he would prefer to work

on. Rather than answer directly, he digressed. "It must be remembered," de Rujula told me, "that the two people most responsible for the development of superstrings, that is to say Green and Schwarz, have spent ten to fifteen years systematically working on something that was not fashionable. In fact they were ridiculed by people for their stubborn adherence to it. So when people come and attempt to convince you that one must work on the most fashionable subject, it is pertinent to remember that the great steps are always made by those who don't work on the most fashionable subject."[4]

I once discussed this situation with a department chair at a major university who lamented that he had not been able to persuade his colleagues to hire John Schwarz in the early 1980s. "They agreed that he was an incredibly smart theorist," he said, "but I couldn't convince them, because they said he was too obsessed and he would probably never work on anything else but string theory. These days I can't convince my colleagues to hire anyone who is *not* a string theorist."

I also recall discussing these problems with Abraham Pais, the particle physicist and biographer of Bohr and Einstein. We used to meet sometimes for lunch at Rockefeller University in New York, where he had been a professor and where I once had an office. "There is nothing you can do," Pais told me. "In my day, too, they were all bastards!"

I think Pais was off the point. This is not about people, it is about how we have structured academic decision making. It is about ensuring that the kinds of scientists needed to advance science have the right opportunities.

This system has another crucial consequence for the crisis in physics: People with impressive technical skills and no ideas are chosen over people with ideas of their own partly because there is simply no way to rank young people who think for themselves. The system is set up not just to do normal science *but to ensure that normal science is what is done.* This was made explicit to me when I applied for my first job out of graduate school. One day, as we were waiting for the results of our applications, a friend came to me looking very worried. A senior colleague had asked him to tell me that I was unlikely to get any jobs, because it was impossible to compare me with other people. If I wanted a career, I had to stop working on my own ideas and work on what other people were doing, because only then could they rank me against my peers.

I can't remember what I thought about this or why it didn't make me crazy with worry. I did have to wait two months longer than everyone else to get a job offer and this was not fun. I was already thinking about what I could do to support myself that wouldn't take too much time from my research. But then I got lucky. The theoretical physics institute in Santa Barbara had just opened, and it had a program in quantum gravity. So my career didn't come to an end.

But it is only now that I understand what was really happening. No one was consciously acting unethically. My friend and his mentor had my best interests, as they understood them, in mind. But here is how a sociologist would describe what happened. The older colleague who gave my friend the message had pioneered a new research program that required challenging calculations. That program needed clever, fast young theorists. I was being told that if I worked on his program, he would give me the gift of a career. It was the simplest and oldest trade in the world: A worker is given the chance to survive in exchange for his labor.

There are many ways the trade is offered, with the takers rewarded and the rebels — those who prefer their own ideas to the ideas of their elders — punished. My friend Carlo Rovelli wanted a job in Rome. He was told to go see a certain professor, who was very friendly and explained to Carlo all about the exciting research program he and his group were pursuing. Carlo thanked him for the insightful discussion and reciprocated by describing his own research program to the professor. The interview was terminated shortly thereafter, and Carlo never got the expected job offer. I had to explain to him what had happened. He was, as we all were once, still naïve enough to think that people were rewarded for having good ideas of their own.

In fact, what it took for Carlo to get a job offer from the University of Rome was to become the leading European scientist in his field. Only then — once he had made an influential career elsewhere, only after hundreds of people around the world had begun to work on his ideas — would the leading professors in Rome be willing to listen to the ideas that he had tried, as a new PhD, to bring to them.

You may wonder how it was that Carlo got a job in the first place. I'll tell you. At that time, in the late 1980s, the field of general relativity was dominated by a few older people who were students of students of Einstein, and they held strongly to the view that the young people with the best and most independent ideas should be encouraged. They led what was called then the relativity community, which had research groups in about a dozen universities in the United States. As a field they were hardly dominant, but they did control a small number of positions, perhaps one new faculty position every two or three years. Carlo was a postdoc in Rome, but due to some bureaucratic problems, his job had never been made official and he had never been paid. Every month he was told that after one more meeting or one more bit of paperwork, he would get a check. After a year and a half of this, he called his friends in the United States and said that although he had not wanted to leave Italy, he was fed up. Were there any jobs available in the United States? As it happened, one of the centers of relativity was looking for an assistant professor, and when they heard he might apply, they flew him over and rushed the appointment through in a matter of weeks. It is worth mentioning that no one at that center worked on quantum gravity — they hired Carlo because he had proved to have original and important ideas in that field.

Could this happen today? Not likely, because now even the field of relativity is dominated by a large research program with a precise agenda set by senior scientists. This has to do with experimental gravitational-wave astronomy and with the hope (still a hope after many years) of doing computer calculations to predict what those experiments should see. These days, a young specialist in general relativity or quantum gravity who is not primarily working on these problems is unlikely to get hired anywhere in the United States.

No matter what the field, a taste of success is often all that is required to turn erstwhile rebels into conservative guardians of their research programs. More than once I've been brought up short by people in my own field of quantum gravity, for supporting the hiring of someone from outside the field who offers new ideas over the hiring of a technically impressive candidate who is working on narrow issues that advance the existing research.

There are really two issues here, and it is important to separate them. One is the dominance of decision making by senior scientists, who frequently use their power to support research programs they invented when they were young and imaginative. The second is the kind of scientist that universities are interested in and able to hire. Do they hire people doing work that everyone in a particular field understands and can judge? Or are they willing to hire people who invent their own directions, which are likely to be difficult to grasp?

This is related to the issue of risk. Good scientists tend to elicit two kinds of responses from referees. Normal, low-risk scientists generally elicit uniform responses; everyone feels the same about them. High-risk scientists and visionaries tend to provoke strongly polarized reactions. There are some people who believe in them deeply and communicate it strongly. Others are highly critical.

The same thing happens when students evaluate their teachers. There is a certain kind of good teacher that students do not feel neutral about. Some love her, or him, and will say, "This is the best teacher I've ever had; this is why I went to college." But others are angry and resentful and hold back nothing on the evaluation form. If you average the scores — reducing the data to a single number, as is often done in deciding the size of professors' raises and chances for tenure — you miss this crucial fact.

Over the years, I've noticed that a polarized distribution of responses is a strong predictor of future success and influence as a scientist. If some people think X is the future of science and others think X is a disaster, this may mean that X is the real thing, someone who aggressively pushes his or her own ideas and has the talent and perseverance to back them up. An environment that embraces risk takers will welcome such people, but a risk-averse environment will shun them.

As far as U.S. research universities are concerned, the basic fact is that people with a polarized set of recommendations frequently don't get hired. Although I have observed this only in my field, it may be generally true. Consider the following scientists, all of them widely admired for the boldness and originality of their contributions to our understanding of evolution: Per Bak, Stuart Kauffman, Lynn Margulis, Maya Paczuski, Robert Trivers. Two of them are physicists who have studied mathematical models of natural selection, the others are leading evolutionary theorists. None made their careers at the most elite universities. When I was younger, I used to wonder why. After a while, I realized that they were too intellectually independent. Their profiles were too bimodal: If many admired them, there were also powerfully placed academics who were skeptical of them. And indeed, it is often the case that the kinds of people who originate ideas are not without faults when measured against the criteria that normal scientists use to judge excellence. They can be too bold. They can be sloppy about the details and unimpressive technically. These criticisms often apply to original thinkers whose curiosity and independence led them into fields they were not trained in. No matter how original and useful their insights, their work will be technically unimpressive to specialists in the domain.

It is also true that some of these creative and original scientists are not easy to get along with. They can be impatient. They express themselves too directly when they disagree with you, and they lack the good manners that come easily to those for whom fitting in is more important than being right. Having known several such "difficult" people, I suspect they are angry for the same reason that very smart women in science are sometimes angry: They have suffered a lifetime of being made to feel marginal.

These kinds of problems certainly affected the career of Per Bak, who, tragically, died of cancer a few years ago, at age fifty-four. He had the rare asset of having written papers in several fields outside his specialty, from economics to cosmology to biology. That should have made him a hot property, in demand at the best universities, but the opposite was the case, because he was not shy to point out that his way of approaching a problem led to insights the experts had missed. He would have had a much better career had he applied his creativity in only one field, but then he wouldn't have been Per Bak.

You might wonder why all the smart people who have become department heads and deans haven't figured all this out and used this knowledge for the benefit of their university. Of course, a few have, and those few will hire such people. Most of the jobs that have opened up in the last decades in non-string approaches to quantum gravity in the United States came about because a chair had a rare opportunity to hire in a field not already represented at his university. Freed from the usual department politics, he made a cost-benefit calculation that convinced him that by hiring in an undersupported field he could instantly obtain a top-ranked group that would raise the status of his department.

Indeed, the problems we are talking about affect all science, and a few influential senior scientists in other fields have expressed their concern. Bruce Alberts is a biologist and past president of the National Academy of Sciences, the most prestigious and influential organization of scientists in the United States. In his presidential address to the National Academy in April 2003, he noted:

> We have developed an incentive system for young scientists that is much too risk-averse. In many ways, we are our own worst enemies. The study sections that we establish to review requests for grant funds are composed of peers who claim that they admire scientific risk-taking, but who generally invest in safe science when allocating resources. The damping effect on innovation is enormous, because our research universities look for assistant professors who can be assured of grant funding when they select new faculty appointments. This helps to explain why so many of our best young people are doing "me too" science.

He went on to describe a trend in which, over the ten years beginning in 1991, the proportion of National Institutes of Health funding going to researchers under thirty-five more than halved, while that going to those over fifty-five increased by more than 50 percent. He bemoaned the result, because it was greatly decreasing the intellectual independence of the younger researchers:

> Many of my colleagues and I were awarded our first independent funding when we were under thirty years old. We did not have preliminary results, because we were trying something completely new. [Now] almost no one finds it possible to start an independent scientific career under the age of thirty-five. Moreover, whereas in 1991 one-third of the principal investigators with NIH funds were under forty, by the year 2002 this fraction had dropped to one-sixth. Even the most talented of our young people seem to be forced to endure several years of rejected grant applications before they finally acquire enough "preliminary data" to assure the reviewers that they are likely to accomplish their stated goals.

Why, if the problem is so obvious that it worries the most prominent leaders of American science, is nothing done about it? This puzzled me for a long time. I now realize that the competition to get and keep academic positions is about more than merit. The system tries to choose the best, most productive people, and to some extent it does that. But there are other agendas, and it would be naïve to ignore them. Equally important, these decisions are about establishing and enforcing a consensus within each field.

Hiring is not the only means of accomplishing this consensus. Everything I've said about hiring is true also of the panels that evaluate grant applications. It also holds for tenure evaluations. These matters are related, because you cannot get tenure in science at a U.S. research university if you haven't been successful in getting grants, and you can't get hired unless there is a likelihood that you will get grants.

As it happened, a few months before I first figured that out, I had been asked to write a piece on another subject for the *Chronicle of Higher Education,* which is a trade journal for university administrators. I wrote to the editors and proposed instead a piece on threats to academic freedom coming from the dominance of popular research programs. They were willing to look at it, but they rejected it as soon as they read my draft. I was outraged: They were suppressing dissent! So I wrote them an unusually (for me) unpleasant e-mail questioning their decision. They responded right away, telling me that the problem was not that the piece was radical — quite the opposite. Everything in it was well known and had been thoroughly aired, within the social sciences and the humanities. They sent me a pile of articles they had published in past years about power relations in academic decision making. I read them and quickly realized that it was only scientists who seem to be ignorant of these issues.

Obviously, there are good reasons to have tenure. To a limited extent, it protects scientists who are original and independent from being fired and replaced by young careerists following the latest intellectual fad. But we pay a severe price for the tenure system: *Too much job security, too much power, and too little accountability for older people. Too little job security, too little power, and too much accountability for younger people in the prime of their creative, risk-taking years.*

Although tenure protects people who are intellectually independent, it does not produce them. I've heard many colleagues say they
are working on what is trendy in order to get tenure, after which
they will do what they really want. But it doesn't seem ever to turn
out that way. I know of only one case where that happened. In the
others, it seems that if these people did not have enough courage
and independence to work on what they wanted when they were
worried about tenure, they did not suddenly gain courage and independence when contemplating what the panel reviewing their grant
would decide. It does not help to have a system that protects
tenured professors' intellectual independence if the same system
makes it unlikely that people with that independence will ever get
tenure.

In fact, professors with tenure who lose their grant funding because of having switched to a more risky area can quickly find
themselves in hot water. They cannot be fired, but they can be pres-
sured with threats of heavy teaching and salary cuts to either go
back to their low-risk, well-funded work or take early retirement.

Here is what Isador Singer, a gifted mathematics professor at
MIT, said recently about the state of his discipline:

> I observe a trend towards early specialization driven by economic
> considerations. You must show early promise to get good letters of
> recommendation to get good first jobs. You can't afford to branch
> out until you have established yourself and have a secure position.
> The realities of life force a narrowness in perspective that is not in-
> herent to mathematics. We can counter too much specialization
> with new resources that would give young people more freedom than
> they presently have, freedom to explore mathematics more broadly,
> or to explore connections with other subjects, like biology these days
> where there is lots to be discovered.
>
> When I was young the job market was good. It was important to
> be at a major university but you could still prosper at a smaller one.
> I am distressed by the coercive effect of today's job market. Young
> mathematicians should have the freedom of choice we had when we
> were young.[5]

The French mathematician Alain Connes has similar criticisms:

The constant pressure [in the U.S. system] for producing reduces the "time unit" of most young people there. Beginners have little choice but to find an adviser [who] is sociologically well implanted (so that at a later stage he or she will be able to write the relevant recommendation letters and get a position for the student) and then write a technical thesis showing that they have good muscles, and all this in a limited amount of time which prevents them from learning stuff that requires several years of hard work. We badly need good technicians, of course, but it is only a fraction of what generates progress in research. . . . From my point of view the actual system in the U.S. really discourages people who are truly original thinkers, which often goes with a slow maturation at the technical level. Also the way the young people get their position on the market creates "feudalities," namely a few fields well implanted in key universities which reproduce themselves leaving no room for new fields. . . . The result is that there are very few subjects which are emphasized and keep producing students and of course this does not create the right conditions for new fields to emerge.[6]

In recent decades, the business world has learned that hierarchy is too costly and has moved to give young people more power and scope. There are now young bankers, software engineers, and the like still in their late twenties who are leading big projects. Every once in a while, you will encounter a lucky academic scientist in the same situation, but it's rare. Many scientists are pushing thirty-five before they emerge from the enforced infancy of a postdoctoral position.

The leaders of hi-tech companies know that if you want to hire the best young engineers, you need young managers. The same holds for other creative fields, such as the music business. I'm sure that some jazz musicians and old rock 'n' roll guys appreciate hip-hop and techno, but the music industry does not let sixty-year-old former stars choose which young musicians get signed to recording contracts. Innovation in music proceeds at such a hectic, vibrant pace because young musicians can find ways to connect to audiences and other musicians quickly, in clubs and on the radio, without having to ask the permission of established artists with their own agendas.

It is interesting to note that the quantum-mechanics revolution was made by a virtually orphaned generation of scientists. Many members of the generation above them had been slaughtered in World War I. There simply weren't many senior scientists around to tell them they were crazy. Today, for graduate students and postdocs to survive, they have to do things that people near retirement can understand. Doing science this way is like driving with the emergency brake on.

Science requires a balance between rebellion and respect, so there will always be arguments between radicals and conservatives. But there is no balance in the current academic world. More than at any time in the history of science, the cards are stacked against the revolutionary. Such people are simply not tolerated in the research universities. Little wonder, then, that even when the science clearly calls for one, we can't seem to pull off a revolution.

20

What We Can Do for Science

I HAVE ATTEMPTED in this book to explain why the list of the five big problems in physics is exactly the same as it was thirty years ago. To tell this story, I have had to focus on string theory, but I want to reiterate that my aim is not to demonize it. String theory is a powerful, well-motivated idea and deserves much of the work that has been devoted to it. If it has so far failed, the principal reason is that its intrinsic flaws are closely tied to its strengths — and, of course, the story is unfinished, since string theory may well turn out to be part of the truth. The real question is not why we have expended so much energy on string theory but why we haven't expended nearly enough on alternative approaches.

When I faced the choice of working on the foundations of quantum mechanics and sabotaging my career or working on a topic related to particle physics so that I might have one, there was a scientific argument bolstering that economic decision. It was obvious that in the previous decades much more progress had been made in particle physics than in plumbing the foundations of quantum theory. A new graduate student today is in a very different situation. The tables have turned. The previous decades have seen little progress in particle theory but a lot in the field of foundations, spurred on by the work in quantum computing.

It has become clear by now that we cannot resolve the five big problems unless we think hard about the foundations of our understanding of space, time, and the quantum. Nor can we succeed if we treat decades-old research programs, like string theory and loop quantum gravity, as if they were established paradigms. We need young scientists with the courage, imagination, and conceptual depth to forge new directions. How do we identify and support such people, instead of discouraging them, as we do now?

I must emphasize again that I do not believe that any individual physicists are to blame for the stasis that has overtaken theoretical physics. Many of the string theorists I know are very good scientists. They have done very good work. I'm not arguing that they should have done better, only that it's remarkable that a large number of the best among us have not been able to succeed, given what seemed at the beginning to be such a good idea.

What we are dealing with is a sociological phenomenon in the world of academic science. I do think that the ethics of science have been to some degree corrupted by the kind of groupthink explored in chapter 16, but not solely by the string theory community. For one thing, it is the academic community writ large that makes the rules. In a court of law, a good lawyer will do anything within the law to advance the cause of his clients. We should expect that the leaders of a scientific field will likewise do everything within the unwritten rules of academia to advance their research program. If the result is the premature takeover of a field by an aggressively promoted set of ideas that achieves its dominance by promising more than it delivers, this cannot be blamed only on its leaders, who are simply doing their jobs based on their understanding of how science works. It can and should be blamed on all of us academic scientists, who collectively make the rules and evaluate the claims made by our colleagues.

It is perhaps a lot to ask that all of us in a field check every result before accepting it as established fact. We can and do leave that to experts in subfields. But it is our responsibility to at least keep track of the claims and the evidence. As much as any of my colleagues, I have been at fault for accepting widely held beliefs about string theory despite their having no support in the scientific literature.

The right question to ask, then, is: What has happened to the traditional constraints of scientific ethics? As we have seen, there is a problem with the structure of academic science, as manifested in such practices as peer review and the tenure system. This is partly responsible for the dominance of string theory, but equally at fault is the confusion of normal science with revolutionary science. String theory began as an attempt to do revolutionary science, yet it is treated as if it were another research program within normal science.

A few chapters back, I suggested that there are two kinds of theoretical physicists, the master craftspeople who power normal science and the visionaries, the seers, who can see through unjustified but universally held assumptions and ask new questions. It should be abundantly clear by now that to make a revolution in science, we need more of the latter. But, as we have seen, these people have been marginalized if not excluded outright from the academy, and they are no longer considered, as they once were, a part of mainstream theoretical physics. If our generation of theorists has failed to make a revolution, it is because we have organized the academy in such a way that we have few revolutionaries, and most of us don't listen to the few we have.

I have concluded that we must do two things. We must recognize and fight the symptoms of groupthink, and we must open the doors to a wide range of independent thinkers, being sure to make room for the peculiar characters needed to make a revolution. A great deal rests on how we treat the next generation. To keep science healthy, young scientists should be hired and promoted based only on their ability, creativity, and independence, without regard to whether they contribute to string theory or any other established research program. People who invent and develop their own research programs should even be given priority, so that they can have the intellectual freedom to work on the approach they judge the most promising. The governance of science is always a matter of making choices. To prevent overinvestment in speculative directions that may turn out to be dead ends, physics departments should ensure that rival research programs and different points of view toward unsolved problems are represented on their faculties — not only because most of the time we cannot predict which views will be right but because the friendly rivalry between smart people working in close proximity is often a source of new ideas and directions.

An openly critical and candid attitude should be encouraged. People should be penalized for doing superficial work that ignores hard problems and rewarded for attacking the long-standing open conjectures, even if progress takes many years. More room could be made for people who think deeply and carefully about the foundational issues raised by attempts to unify our understanding of space and time with quantum theory.

Many of the sociological problems we've discussed have to do with the tendency of scientists — indeed, all human beings — to form tribes. To combat this tribal tendency, string theorists could de-emphasize boundaries between string theory and other approaches. They could stop categorizing theorists by whether or not they display loyalty to this or that conjecture. People who work on alternatives to string theory, or who are critical of string theory, should be invited to speak at string theory conferences, to the benefit of all. Research groups should seek out postdocs, students, and visitors who pursue rival approaches. Students should also be encouraged to learn about and work on competing approaches to unsolved problems, so that they are equipped to choose for themselves the most promising directions as their careers advance.

We physicists need to confront the crisis facing us. A scientific theory that makes no predictions and therefore is not subject to experiment can never fail, but such a theory can never succeed either, as long as science stands for knowledge gained from rational argument borne out by evidence. There needs to be an honest evaluation of the wisdom of sticking to a research program that has failed after decades to find grounding in either experimental results or precise mathematical formulation. String theorists need to face the possibility that they will turn out to have been wrong and others right.

Finally, there are many steps that organizations supporting science can take to keep science healthy. Funding agencies and foundations should enable scientists at every level to explore and develop viable proposals to solve the deep and difficult problems. A research program should not be allowed to become institutionally dominant before it has gathered convincing scientific proof. Until it does, alternative approaches should be encouraged, so that the progress of science is not stalled by overinvestment in a wrong direction. When there is a recalcitrant but key problem, there should be a limit on the proportion of support given to any one research program that aims to solve it — say, a third of total funding.

Some of these proposals represent major reforms. But when it comes to theoretical physics, we are not talking about much money at all. Suppose that an agency or foundation decided to fully support *all* the visionaries who ignore the mainstream and follow their own ambitious programs to solve the problems of quantum gravity and quantum theory. We are talking about perhaps two dozen theorists. Supporting them fully would take a tiny fraction of any large nation's budget for physics. But judging from what such people have contributed in the past, it's likely that several will do something important enough to make this the best overall investment in the field.

Indeed, even small foundations can help, by searching out independent-minded seers with PhDs in theoretical physics or mathematics who are working on their own approach to a fundamental problem — people, that is, doing something so unconventional that there's a good chance they will never be able to have an academic career. Someone like Julian Barbour, Antony Valentini, Alexander Grothendieck — or Einstein, for that matter. Give them five years of support, extendable to a second or even third five-year term if they are getting anywhere at all.

Sound risky? The Royal Society in the United Kingdom has a program like this. It has been responsible for jump-starting the careers of several scientists who are now important in their field and who would probably never have gotten that sort of support in the United States.

How would you choose those people worthy of support? Simple. Ask some who already do science this way. Just to be sure, find at least one accomplished person in the candidate's field who is deeply excited about what the candidate is trying to do. To be really sure, find at least one professor who thinks the candidate is a terrible scientist and bound to fail.

It may seem strange to be discussing academic politics in a book for the general public, but you, the public, individually and collectively, are our patrons. If the science you pay for is not getting done, it is up to you to hold our feet to the fire and make us do our job.

So I have some final words for different audiences.

To the educated public: Be critical. Don't believe most of what you hear. When a scientist claims to have done something impor- tant, ask to see the evidence. Evaluate it as strictly as you would an investment. Give it as much scrutiny as a house you would buy or a school you would send your children to.

To those who make decisions about what science gets done — that is, to department heads, search committees, deans, officers of foundations, and funding agencies: Only people at your level can implement recommendations like the ones just listed. Why not con- sider them? These are proposals that should be discussed in places like the offices of the National Science Foundation, the National Academy of Sciences, and their counterparts around the world. This is not just a problem for theoretical physics. If a highly disciplined subject like physics is vulnerable to the symptoms of groupthink, what may be happening in other, less rigorous areas?

To my fellow theoretical physicists: The problems discussed in this book are the responsibility of all of us. We constitute a scien- tific elite only because the larger society we are part of cares deeply about the truth. If string theory is wrong but continues to dominate our field, the consequences could be severe — for us personally as well as for our profession. It's up to us to open the doors and allow the alternatives in, and generally raise the standards of argument.

To put it more bluntly: If you are someone whose first reaction when challenged on your scientific beliefs is "What does X think?" or "How can you say that? Everybody good knows that . . . ," then you are in danger of no longer being a scientist. You are paid good money to do your job, and that means you have a responsibility to make a careful and independent evaluation of everything you and your colleagues believe. If you cannot give a precise defense of your beliefs and commitments, consistent with the evidence, if you let other people do your thinking for you (even if they are senior and powerful), then you are not living up to your ethical obligations as a member of a scientific community. Your doctorate is a license for you to hold your own views and make your own judgments. But it is more than that; it obliges you to think critically and indepen- dently about everything in your domain of competence.

This is harsh. But here are even harsher words for those of us working on fundamental problems who are not string theorists. Our job is supposed to be to find the wrong assumptions, ask the new questions, find the new answers, and lead revolutions. It's easy to see where string theory is probably wrong, but criticizing string theory is not the job. The job is to invent the theory that's right.

I'll be harshest on myself. I fully expect some readers to come back at me with "If you're so smart, why haven't you done any better than the string theorists?" And they'd be right. Because in the end, this book is a form of procrastination. Of course, I hope by writing it to make the way easier for those who will follow. But my craft is theoretical physics and my real job is to finish the revolution Einstein started. I haven't done that job.

So what am I going to do myself? I'm going to try to take advantage of the good fortune that life has shown me. To begin with, I'm going to dig out my old paper, "On the Relationship Between Quantum and Thermal Fluctuations," and read it. Then I'm going to turn off the phone and the BlackBerry, put on some Bebel Gilberto, Esthero, and Ron Sexsmith, turn the volume way up, erase the blackboard, get out some good chalk, open a new notebook, take out my favorite pen, sit down, and start thinking.

Notes
Acknowledgments
Index

Notes

Introduction

1. Mark Wise, "Modifications to the Properties of the Higgs Boson," Seminar talk, Mar. 23, 2006. Available at http://streamer.perimeterinstitute.ca:81/mediasite/.

2. Brian Greene, *The Fabric of the Cosmos: Space, Time, and the Texture of Reality* (New York: Alfred A. Knopf, 2005), p. 376.

3. Gerard 't Hooft, *In Search of the Ultimate Building Blocks* (Cambridge: Cambridge University Press, 1996), p. 163.

4. Quoted in *New Scientist*, "Nobel Laureate Admits String Theory Is in Trouble," Dec. 10, 2005. This raised something of a controversy, so Gross clarified his remarks at the opening of the 23rd Jerusalem Winter School in Theoretical Physics (full text available at www.as.huji.ac.il/schools/phys23/media.shtml):

> What I really meant by that is that we still don't know the answer as to what string theory is and whether it's the final theory or there is something missing in it and we seem to be faced with the necessity of deep conceptual changes . . . especially as to the nature of space and time. But, far from being an argument that we should stop doing string theory — it's failed, it's over — this is a wonderful period.

5. J. Polchinski, talk given at the 26th SLAC Summer Institute on Particle Physics, 1998, hep-th/9812104.

6. http://motls.blogspot.com/2005/09/why-no-new-einstein-ii.html.

7. Lisa Randall, "Designing Words," in *Intelligent Thought: Science Versus the Intelligent Design Movement*, ed. John Brockman (New York: Vintage, 2006).

1. The Five Great Problems in Theoretical Physics

1. John Stachel, "How Did Einstein Discover Relativity?" http://www .aip.org/history/einstein/essay-einstein-relativity.htm. I should note that some philosophers of science regard general relativity as at least partly a constructive theory; for the purposes of this discussion it is a principle theory because it describes how space, time, and motion are to be described, whatever matter the universe contains.

2. The Beauty Myth

1. The reader should note that my telling of this story simplifies it a great deal to make a point. There were other crucial experiments, which had to do with light moving through flowing water or the effect on observations of starlight of the relative motion of Earth and the star. Einstein was also not the only one who realized that the right answer involved embracing the principle of relativity; so did the great French mathematician and physicist Henri Poincaré.

3. The World As Geometry

1. I should confess that this is not how Nordström solved the problem. But he might have. This was the way adopted by later proponents of extra dimensions and it is an improvement over what Nordström did.

2. There is a caveat, which is that this applies only to observations that take place in small regions of space over small intervals of time. If you fall far enough to see that the strength of the gravitational field changes, you can distinguish gravity from acceleration.

3. An expert might prefer the more precise notion of inertia here, but I have found that this confuses lay readers.

4. Except of course, in the case of dark matter and dark energy, as we have seen.

5. Quoted in Hubert F. M. Goenner, *On the History of Unified Field Theories (1914–1933)*, p. 30. http://relativity.livingreviews.org/Articles/lrr-2004-2/index.html (2004).

6. Ibid., pp. 38–39.

7. Ibid., p. 39.

8. Ibid., p. 35.

9. Quoted in Abraham Pais, *Subtle Is the Lord* (New York: Oxford Univ. Press, 1982), p. 330.

10. Ibid., p. 332.

11. Ibid.

12. Ibid., p. 334.

4. Unification Becomes a Science

1. Those readers who are interested to learn more may read about gauge symmetries in chapter 4 of my 1997 book, *The Life of the Cosmos* (New York: Oxford University Press).

2. Although we won't need it, some readers may want to know more about how the gauge principle works. Here is the key idea: Usually the operations that define a symmetry have to be done to the whole system. To show that an object is symmetric under rotation, you have to rotate the whole object at once. You can't rotate only part of a ball. But there are special cases in which the symmetry works even if you apply it to a part of the system. Such symmetries are called *local* symmetries. This seems counterintuitive; how could it work? It turns out — and this is the hard thing to explain without math — that it works if the various parts of the system act on one another with certain forces. These are the gauge forces.

3. Again, the history is more complicated than my summary. The Yang-Mills theories were actually first discovered in the 1920s context of the higher-dimensional unified theories but appear to have been forgotten, leading to their rediscovery by Chen Ning Yang, Robert Mills, and others in the 1950s.

4. The main theme of *The Life of the Cosmos* was the implications of this change.

5. From Unification to Superunification

1. Y. Nomura and B. Tweedie, hep-ph/0504246.

2. P. Frampton, e-mail (used with permission).

6. Quantum Gravity: The Fork in the Road

1. Einstein, "Approximate Integration of the Field Equations of Gravitation," *Sitzungsberichte der Preussische Akadamie der Wissenschaften* (Berlin, 1916), pp. 688–96. For the early history of quantum gravity, see John Stachel, introduction and comments to Part V, *Conceptual Foundations of Quantum Field Theory*, ed. Tian Yu Cao (Cambridge, U.K.: Cambridge University Press, 1999).

2. W. Heisenberg and W. Pauli, "Zur Quantendynamik der Wellenfelder," *Zeit. für Physik*, 56:1–61 (1929), p. 3.

3. M. P. Bronstein, "Quantization of gravitational waves," *Zh. Eksp. Teor. Fix.* 6 (1936), p. 195. For more information about Bronstein, see Stachel in *Conceptual Foundations*, and also G. Gorelik, "Matvei Bronstein and Quantum Gravity: 70th Anniversary of the Unsolved Problem," *Physics-Uspekhi*, 48:10 (2005).

4. Richard P. Feynman, *What Do You Care What Other People Think?* (New York: W. W. Norton, 1988), p. 91.

5. In fact this is a general property of systems bound together by gravity, such as stars and galaxies. All these are systems that cool down when energy is put in. This fundamental difference between systems with and without gravity has proved to be a big stumbling block for many attempts to unify physics.

7. Preparing for a Revolution

1. G. Veneziano, "Construction of a Crossing-Symmetric Regge-Behaved Amplitude for Linearly Rising Regge Trajectories," *Nuovo Cim.*, 57 A:190–97 (1968).

2. http://www.edge.org/3rd_culture/susskind03/susskind_index.html.

3. P. Ramond, "Dual theory for free fermions," *Phys. Rev. D*, 3(10):2415–18 (1971).

4. Another particularly influential paper was "Quantum Dynamics of a Massless Relativistic String," by P. Goddard, J. Goldstone, C. Rebbi, and C. Thorn, *Nucl. Phys.*, 56:109–35 (1973).

5. J. Scherk and J. H. Schwarz, "Dual Models for Non-Hadrons," *Nucl. Phys. B*, 81(1):118–44 (1974).

6. T. Yoneya, "Connection of Dual Models to Electrodynamics and Gravidynamics," *Prog. Theor. Phys.*, 51(6):1907–20 (1974).

8. The First Superstring Revolution

1. J. H. Schwarz, interviewed by Sara Lippincott, July 21 and 26, 2000, http://oralhistories.library.caltech.edu/116/01/Schwarz_OHO.pdf.

2. M. B. Green and J. H. Schwarz, "Anomaly Cancellations in Supersymmetric D=10 Gauge Theory and Superstring Theory," *Phys. Lett. B*, 149 (1–3):117–22 (1984).

3. Schwarz interview.

4. Thomas S. Kuhn, *The Structure of Scientific Revolutions* (Chicago: Univ. of Chicago Press, 1962).

5. S. Mandelstam, "The N-loop String Amplitude — Explicit Formulas, Finiteness and Absence of Ambiguities," *Phys. Lett. B*, 277(1–2):82–88 (1992).

6. P. Candelas et al., "Vacuum Configurations for Superstrings," *Nucl. Phys. B*, 258(1):46–74 (1985).

7. A. Strominger, "Superstrings with Torsion," *Nucl. Phys. B*, 274(2): 253–84 (1986).

8. In P.C.W. Davies and Julian Brown, eds., *Superstrings: A Theory of Everything* (Cambridge, U.K.: Cambridge Univ. Press, 1988), pp. 194–95.

9. Sheldon L. Glashow and Ben Bova, *Interactions: A Journey Through the Mind of a Particle Physicist* (New York: Warner Books, 1988), p. 25.

10. L. Smolin, "Did the Universe Evolve?" *Class. Quant. Grav.*, 9(1): 173–91 (1992).

9. Revolution Number Two

1. E. Witten, "String Theory Dynamics in Various Dimensions," hep-th/9503124; *Nucl. Phys. B*, 443:85–126 (1995).

2. C. M. Hull and P. K. Townsend, "Unity of Superstring Dualities," hep-th/9410167; *Nucl. Phys. B*, 438:109–37 (1994).

3. J. Polchinski, "Dirichlet Branes and Ramond-Ramond Charges," *Phys. Rev. Lett.*, 75(26):4724–27 (1995).

4. J. Maldacena, "The Large N Limit of Superconformal Field Theories and Supergravity," hep-th/9711200; *Adv. Theor. Math. Phys.*, 2:231–52 (1998); *Int. J. Theor. Phys.*, 38:1113–33 (1999).

5. A. M. Polyakov, "A Few Projects in String Theory," hep-th/9304146.

6. B. de Wit, J. Hoppe, and H. Nicolai, "On the Quantum-Mechanics of Supermembranes," *Nucl. Phys. B*, 305(4):545–81 (1988).

7. T. Banks, W. Fischler, S. Shenker, and L. Susskind, "M-Theory as a Matrix Model: A Conjecture," *Phys. Rev. D*, 55(8):5112–28 (1997).

10. A Theory of <u>Anything</u>

1. The supernova observations were made by Saul Perlmutter and collaborators at Lawrence Berkeley Laboratory and Robert Kirschner and colleagues in the High-Z Supernova Search Team.

2. E. Witten, "Quantum Gravity in de Sitter Space," hep-th/0106109. Witten continues, "This last statement is not very surprising given the classical no go theorem. For, in view of the usual problems in stabilizing moduli, it is hard to get de Sitter space in a reliable fashion at the quantum level given that it does not arise classically."

3. S. Kachru, R. Kallosh, A. Linde, and S. Trivedi, "De Sitter Vacua in String Theory," hep-th/0301240.

4. See, for example, T. Hertog, G. T. Horowitz, and K. Maeda, "Negative Energy Density in Calabi-Yau Compatifications," hep-th/0304199, *Jour. High Energy Phys.*, 0305:60(2003).

11. The Anthropic Solution

1. L. Susskind, "The Anthropic Landscape of String Theory," hep-th/0302219.

2. S. Weinberg, "Anthropic Bound on the Cosmological Constant," *Phys. Rev. Lett.*, 59(22):2607–10 (1987).

3. L. Smolin, "Did the Universe Evolve?" *Class. Quant. Grav.*, 9(1): 173–91 (1992).

4. Weinberg, "Living in the Multiverse," hep-th/0511037.

5. From a recent survey by *Seed Magazine* on the relationship between the anthropic principle and the proliferation of string theories; http://www.seedmagazine.com/news/2005/12/surveying_the_landscape.

6. E. J. Copeland, R. C. Myers, and J. Polchinski, "Cosmic F- and D-Strings," *Jour. High Energy Phys.*, Art. no. 013, June 2004.

7. M. Sazhin et al., "CSL-1: Chance Projection Effect or Serendipitous Discovery of a Gravitational Lens Induced by a Cosmic String?" *Mon. Not. R. Astron. Soc.*, 343:353–59 (2003).

8. N. Arkani-Hamed, G. Dvali, and S. Dimopoulos, "The Hierarchy Problem and New Dimensions at a Millimeter," *Phys. Lett. B*, 429:263–72 (1998).

9. L. Randall and R. Sundrum, "An Alternative to Compactification," hep-th/9906064; *Phys. Rev. Lett.*, 83:4690–93 (1999).

12. What String Theory Explains

1. In technical terms, supersymmetry implies that there is a timelike or lightlike killing field on the spacetime geometry. It implies the existence of a symmetry in time, because (in technical language) the supersymmetry algebra closes on the Hamiltonian. Another way to say this is that supersymmetry requires a killing spinor, which implies a null or timelike killing vector.

2. E. D'Hoker and D. H. Phong, *Phys. Lett. B*, 529:241–55 (2002); hep-th/0110247.

3. D. Friedan, "A Tentative Theory of Large Distance Physics," hep-th/0204131.

4. D. Karabali, C. Kim, and V. P. Nair, *Phys. Lett. B*, 434:103–9 (1998); hep-th/9804132; R. G. Leigh, D. Minic, and A. Yelnikov, hep-th/0604060. For the application to 3+1 dimensions, see L. Freidel, hep-th/0604185.

5. In *The Road to Reality* (2005), Roger Penrose argued that most of the compactified spaces that extra dimensions curl up into will quickly collapse to singularities. To show this, he applied to the spacetime backgrounds of these string theories the theorems he and Hawking developed showing that general relativity predicts singularities in cosmological solutions. As far as I know, his arguments stand. They hold only at the classical level of approximation, but this is the only approximation in which we can study the time evolution of spacetime backgrounds in string theory. Therefore, Penrose's result is as reliable as the arguments that convince string theorists of the existence of the landscape of string theories.

6. Quoted in Amanda Gefter, "Is String Theory in Trouble?" *New Scientist*, Dec. 17, 2005.

13. Surprises from the Real World

1. It is often the case that surprising experimental results are not confirmed when other experimentalists repeat the experiment. This does not mean someone is being dishonest. Experiments on the edge of what is possible are almost always hard to replicate, and it is typically difficult to separate noise from a meaningful signal. Often it takes many years and many attempts by different people before all the sources of error in a new kind of experiment are understood and eliminated.

2. Expressed in terms of R, the cosmological constant is equal to $1/R^2$.

3. K. Land and J. Magueijo, "Examination of Evidence for a Preferred Axis in the Cosmic Radiation Anisotropy," *Phys. Rev. Lett.*, 95:071301 (2005).

4. Ibid.

5. M. Milgrom, "A Modification of the Newtonian Dynamics as a Possible Alternative to the Hidden Mass Hypothesis," *Astrophys. Jour.*, 270(2): 365–89 (1983).

6. More information about MOND and the data supporting it, plus references, is available at www.astro.umd.edu/~ssm/mond/.

7. J. D. Anderson et al., "Study of the Anomalous Acceleration of Pioneer 10 and 11," gr-qc/0104064.

8. M. T. Murphy et al., "Further Evidence for a Variable Fine Structure Constant from Keck/HIRES QSO Absorption Spectra," *Mon. Not. Roy. Ast. Soc.*, 345:609–38 (2003).

9. See, for example, E. Peik et al., "Limit on the Present Temporal Variation of the Fine Structure Constant," *Phys. Rev. Lett.*, 93(17):170801 (2004), and R. Srianand et al., "Limits on the Time Variation of the Electromagnetic Fine Structure Constant in the Low Energy Limit from Absorption Lines in the Spectra of Distant Quasars," *Phys. Rev. Lett.*, 92(12):121302 (2004).

10. K. Greisen, "End to the Cosmic-Ray Spectrum?" *Phys. Rev. Lett.*, 16(17):748–50 (1966), and G. T. Zatsepin and V. A. Kuzmin, "Upper Limit of the Spectrum of Cosmic Rays," *JETP Letters*, 4:78–80 (1966).

11. S. Coleman and S. L. Glashow, "Cosmic Ray and Neutrino Tests of Special Relativity," *Phys. Rev. B*, 405:249–52 (1997); Coleman and Glashow, "Evading the GZK Cosmic-Ray Cutoff," hep-ph/9808446.

14. Building on Einstein

1. G. Amelino-Camelia, "Testable Scenario for Relativity with Minimum-Length," hep-th/0012238.

2. João Magueijo, *Faster Than the Speed of Light: The Story of a Scientific Speculation* (New York: Perseus Books, 2003).

3. Vladimir Fock, *The Theory of Space, Time, and Gravitation* (London: Pergamon Press, 1959).

4. L. Friedel, J. Kowalski-Glikman, and L. Smolin, "2 + 1 Gravity and Doubly Special Relativity," *Phys. Rev. D.*, 69:044001 (2004).

5. E. Livine and L. Friedel, "Ponzano-Regge Model Revisited III: Feynman Diagrams and Effective Field Theory," hep-th/0502106; *Class. Quant. Grav.*, 23:2021–62 (2006).

6. Florian Girelli and Etera R. Livine, "Physics of Deformed Special Relativity," gr-qc/0412079.

15. Physics After String Theory

1. A. Ashtekar, "New Variables for Classical and Quantum Gravity," *Phys. Rev. Lett.*, 57(18):2244–47 (1986).

2. http://online.kitp.ucsb.edu/online/kitp25/witten/oh/10.html.

3. This was not always the prevalent belief; credit for championing the role of causality should go to Roger Penrose, Rafael Sorkin, Fay Dowker, and Fotini Markopoulou.

4. See, for example, R. Loll, J. Ambjørn, and J. Jurkiewicz, "The Universe from Scratch," hep-th/0509010.

5. See, for example, Alain Connes, *Noncommutative Geometry* (San Diego: Academic Press, 1994).

6. O. Dreyer, "Background-Independent Quantum Field Theory and the Cosmological Constant Problem," hep-th/0409048.

7. See, for example, C. Rovelli, "Graviton Propagator from Background-Independent Quantum Gravity," gr-qc/0508124.

8. S. Hofmann and O. Winkler, "The Spectrum of Fluctuations in Singularity-free Inflationary Quantum Cosmology," astro-ph/0411124.

9. F. Markopoulou, "Towards gravity from the quantum," hep-th/0604120.

10. S. O. Bilson-Thompson, "A Topological Model of Composite Preons," hep-ph/0503213.

11. S. O. Bilson-Thompson, F. Markopoulou, and L. Smolin, "Quantum Gravity and the Standard Model," hep-th/0603022.

12. Audiotapes of the discussions are available at http://www.perimeterinstitute.ca/activities/scientific/cws/evolving_laws/.

16. How Do You Fight Sociology?

1. www.cosmicvariance.com/2005/11/18/a-particle-physicists-perspective.

2. Anonymous posting on http://groups.google.com/group/sci.physics.strings/by String Theorist, Oct. 9, 2004.

3. *Guardian Unlimited,* Jan. 20, 2005.

4. To papers of mine questioning one or another result in string theory, I've received three responses in which the correspondent refers to the "strong" theory community. As in "while perturbative finiteness (or the Maldacena conjecture, or S-duality) may not have been proved, no one in the strong theory community believes that it could possibly be false." Once might be a coincidence; thrice, and this is a classic Freudian slip. How much of the sociology of string theory is just the all-too-recognizable human desire to want to be part of the strongest group around?

5. S. Kachru, R. Kallosh, A. Linde, and S. Trivedi, "De Sitter Vacua in String Theory," hep-th/0301240.

6. http://groups.google.com/group/sci.physics.strings/, April 6, 2004.

7. L. Smolin, "Did the Universe Evolve?" *Class. Quant. Grav.,* 9:173–91 (1992).

8. www.imp.ac.ir/IPM/news/connes-interview.pdf (used with permission).

9. Michael Duff, *Physics World,* Dec. 2005.

10. www.damtp.cam.ac.uk/user/gr/public/qg_ss.html.

11. However, I am glad to report that it is not hard to find on the Web introductions to string theory that do *not* make distorted or exaggerated claims. Here are some examples: http://tena4.vub.ac.be/beyondstringtheory/index.html; http://www.sukidog.com/jpierre/strings/;http://en.wikipedia.org/wiki/M-theory.

12. S. Mandelstam, "The N-loop String Amplitude—Explicit Formulas, Finiteness and Absence of Ambiguities," *Phys. Lett. B,* 277(1–2):82–88 (1992).

13. Here are a few examples: J. Barbon, hep-th/0404188, *Eur. Phys. J.,* C33: S67–S74 (2004); S. Foerste, hep-th/0110055, *Fortsch. Phys.,* 50:221–403 (2002); S. B. Giddings, hep-th/0501080; and I. Antoniadis and G. Ovarlez, hep-th/9906108. A rare example of a review with a careful and correct (for the time) discussion of the issue of finiteness is L. Alvarez-Gaume and M. A. Vazquez-Mozo, hep-th/9212006.

14. This is a paper by Andrei Marshakov (*Phys. Usp.,* 45:915–54 (2002), hep-th/0212114). I apologize for the technical language, but perhaps the reader can see the point:

> Unfortunately the ten-dimensional superstring pretending to be the most successful among existing string models is strictly defined, in general, only at tree and one-loop levels. Starting from the two-loop corrections to the scattering amplitudes all expressions in the perturbative superstring theory are really not defined. The reason for that comes from the well-known problems with supergeometry or integration over the "superpartners" of the moduli of complex structures. In contrast to the bosonic case where the integration measure is fixed by the Belavin-Knizhnik theorem, the definition of the integration measure over supermoduli (or, more strictly, the odd moduli of super-complex structures) is still an unsolved problem [88, 22]. The moduli

spaces of the complex structures of Riemann surfaces are non compact, and the integration over such spaces requires special care and additional definitions. In the bosonic case, when the integrals over moduli spaces diverge, the result of integration in (3.14) is defined only up to certain "boundary terms" (the contributions of degenerate Riemann surfaces or the surfaces of lower genera (with less "handles"). In the superstring case one runs into more serious problems since the very notion of the "boundary of moduli space" is not defined. Indeed the integral over the Grassmann odd variables does not "know" what is the boundary term. This is the fundamental reason why the integration measure in fermionic string is not well-defined and depends on the "gauge choice" or the particular choice for the "zero modes" in the action (3.23). For two-loop contributions this problem can be solved "empirically" (see [88, 22]), but in the general setup the superstring perturbation theory is not mathematically well defined. Moreover, these are not problems of the formalism: the same obstacles arise in less geometrical approach of Green and Schwarz [91].

15. Here is an e-mail from Mandelstam, dated June 8, 2006:

With regard to my paper on the finiteness of the n-loop string amplitude, let me first remark that divergences can only occur where the moduli space degenerates. I examined the points of degeneracy associated with the "dilaton" divergence, with which string theorists have been concerned. I showed that the arguments previously applied to the one-loop amplitude can be extended to the n-loop amplitude, and also that the associated ambiguities in the definition of the integration contour over the even supermoduli can be resolved by using the unique prescription consistent with unitarity. I agree that this does not provide a mathematically rigorous proof of finiteness, but I believe it deals with the physical problems which could lead to infinities. I did not examine another source of infinities, known from the early days of dual models, namely the use of imaginary time. The factor $exp(iEt)$, where E is the difference between the immediate and initial energies, can clearly diverge if one integrates over imaginary time. One believes on physical grounds that such infinities can be removed by analytic continuation to real time. This has been shown explicitly for the zero- and one-loop amplitude, and it has been shown that an analytic continuation leading to finiteness can be defined for the two-loop amplitude.

16. G. T. Horowitz and J. Polchinski, "Gauge/gravity duality," gr-qc /0602037. To appear in *Towards Quantum Gravity*, ed. Daniele Oriti, Cambridge University Press.

17. http://golem.ph.utexas.edu/~distler/blog/archives/000404.html.

18. Irving Janis, *Victims of Groupthink: A Psychological Study of Foreign-Policy Decisions and Fiascoes* (Boston: Houghton Mifflin, 1972), p. 9. Of course, the phenomenon is much older. John Kenneth Galbraith, the influential economist, called it "conventional wisdom." He meant by this

"opinions that, while not necessarily well founded, are so widely held among the rich and influential that only the rash and foolish will endanger their careers by dissenting from them." (From a book review in the *Financial Times*, Aug. 12, 2004.)

19. Irving Janis, *Crucial Decisions: Leadership in Policymaking and Crisis Management* (New York: Free Press, 1989), p. 60.

20. http://oregonstate.edu/instruct/theory/grpthink.html.

21. Another example is the erroneous proof of the nonexistence of hidden variables in quantum theory, published by John von Neumann in 1932 and widely cited for three decades before the quantum theorist David Bohm found a hidden variables theory.

17. What Is Science?

1. See Paul Feyerabend, *Killing Time: The Autobiography of Paul Feyerabend* (Chicago: Univ. of Chicago Press, 1996).

2. See, for example, Karl Popper, *The Logic of Scientific Discovery* (New York: Routledge, 2002).

3. Thomas S. Kuhn, *The Structure of Scientific Revolutions* (Chicago: Univ. of Chicago Press, 1962).

4. Imre Lakatos, *Proofs and Refutations* (Cambridge, U.K.: Cambridge Univ. Press, 1976).

5. Leonard Susskind, in defending the validity of anthropic reasoning, has labeled its critics *Popperazzi*, for invoking the need for some means of falsification. But it is one thing to accept the critiques of Popper holding that falsification is only part of the story of how science works, and quite another to advocate the acceptance on scientific grounds of a theory that makes no unique or specific predictions by which it might be either falsified or confirmed. In this regard, I am proud to be a Popperazzo.

6. Alexander Marshack, *The Roots of Civilization: The Cognitive Beginnings of Man's First Art, Symbol, and Notation* (New York: McGraw-Hill, 1972).

7. D. H. Wolpert and W. G. Macready, *No Free Lunch Theorems for Search*, Technical Report, Santa Fe Institute, SFI-TR-95-02-010.

8. Richard P. Feynman, "What Is Science?" *The Physics Teacher*, Sept. 1969.

18. Seers and Craftspeople

1. Quoted in Simon Singh, "Even Einstein Had His Off Days," *New York Times*, Jan. 2, 2005.

2. See, for example, Mara Beller, *Quantum Dialogue: The Making of a Revolution* (Chicago: Univ. of Chicago Press, 1999).

3. Thomas S. Kuhn, *The Structure of Scientific Revolutions* (Chicago: Univ. of Chicago Press, 1962).

4. A. Einstein to R. A. Thorton, unpublished letter dated Dec. 7, 1944 (EA 6-574). Einstein Archive, Hebrew University, Jerusalem. Quoted in Don Howard, "Albert Einstein as a Philosopher of Science," *Physics Today*, Dec. 2005.

5. T. Jacobson and L. Smolin, "Nonperturbative Quantum Geometries," *Nucl. Phys. B*, 299:295–345 (1988).

6. See, for example, L. Crane, "Clock and Category: Is Quantum Gravity Algebraic?" gr-qc/9504038; *J. Math. Phys.*, 36:6180–193 (1995).

7. See, for example, F. Markopoulou, "An Insider's Guide to Quantum Causal Histories," hep-th/9912137; *Nucl. Phys. B*, Proc. Supp., 88(1): 308–13 (2000).

8. Seth Lloyd, *Programming the Universe: A Quantum Computer Scientist Takes On the Cosmos* (New York: Alfred A. Knopf, 2006).

9. I have here to again emphasize that I am talking only about people with good training all the way through to a PhD. This is not a discussion about quacks or people who misunderstand what science is.

10. L. Smolin, "On the Nature of Quantum Fluctuations and Their Relation to Gravitation and the Principle of Inertia," *Class. Quant. Grav.*, 3:347–59 (1986).

11. Julian Barbour, *The End of Time: The Next Revolution in Physics* (New York: Oxford Univ. Press, 2001).

12. D. Finkelstein, "Past-Future Asymmetry of the Gravitational Field of a Point Particle," *Phys. Rev.*, 110: 965–67 (1958).

13. Antony Valentini, *Pilot-Wave Theory of Physics and Cosmology* (Cambridge, U.K.: Cambridge Univ. Press, in press).

14. Here is part of a letter from the National Science Foundation to University of Notre Dame physicist James Cushing in 1995, rejecting his proposal to support his work on foundations of quantum theory:

> The subject under consideration, the rival Copenhagen and causal [Bohm] interpretations of the quantum theory, has been discussed for many years and in the opinion of several members of the Physics Division of the NSF, the situation has been settled. The causal interpretation is inconsistent with experiments which test Bell inequalities. Consequently . . . funding . . . a research program in this area would be unwise.

The remarkable thing about this letter is that it contains an elementary mistake, as it was by then well understood by experts that the causal interpretation is fully consistent with the experiments that test the Bell inequalities. By the way, Cushing had been a successful elementary-particle physicist before switching his interests to the foundations of quantum theory, but this did not prevent the NSF from cutting off his funding.

15. D. Deutsch, *Proc. Roy. Soc. A*, 400:97–117 (1985).

16. David Deutsch, *The Fabric of Reality: The Science of Parallel Universes and Its Implications* (London: Penguin, 1997).

17. P. W. Shor, "Polynomial-Time Algorithms for Prime Factorization and Discrete Logarithms on a Quantum Computer," quant-ph/9502807.

18. A. Valentini, "Extreme Test of Quantum Theory with Black Holes," astro-ph/041250s.

19. Alexander Grothendieck, *Récoltes et Semailles*, 1986, English translation by Roy Lisker, www.grothendieck-circle.org, chapter 2.

19. How Science Really Works

1. There is an unfortunate exception to this, which is when a professor is frightened by his younger self and has renounced his youthful risk-taking spirit in favor of scientific conservatism. Reminding someone like this of a younger self is generally not a good idea.

2. See, for example, "A Study on the Status of Women Faculty in Science at MIT," vol. XI, no. 4, March 1999, available online at http://web.mit.edu/fnl/women/women.html. More information on issues on women in science is available from the American Physical Society at http://www.aps.org/educ/cswp/ and from the committee on Faculty Diversity at Harvard University at http://www.aps.org/educ/cswp/.

3. James Glanz, "Even Without Evidence, String Theory Gains Influence," *New York Times*, March 13, 2001.

4. Gary Taubes, *Nobel Dreams: Power, Deceit and the Ultimate Experiment* (New York: Random House, 1986), pp. 254–55.

5. Isador Singer, from an interview published online at http://www.abelprisen.no/en/prisvinnere/2004/interview_2004_1.html.

6. Alain Connes, interview available at www.ipm.ac.ir/IPM/news/connes-interview.pdf.

Acknowledgments

A book starts with an idea, and the credit for this one goes to John Brockman for perceiving that I wanted to do something more than write an obscure academic monograph on the relationship between democracy and science. That is one topic of this book, but, as he foresaw, the argument is much more powerful when developed in the context of a specific scientific controversy. I am indebted to him and to Katinka Matson for their continuing support and for inviting me into the community that comprises the third culture. By offering me a context that goes beyond my specialization, they changed my life.

No writer has had a better editor than Amanda Cook, and the extent to which anything good here is due to her guidance and interventions is embarrassing to admit. Sara Lippincott finished the job with an elegance and precision any writer would kill for. It was an honor to work with both of them. Holly Bemiss, Will Vincent, and everyone at Houghton Mifflin took care of this book with enthusiasm and skill.

Over the last decades, many colleagues have taken their time to educate me about string theory, supersymmetry, and cosmology. Among them, I am especially grateful to Nima Arkani-Hamed, Tom Banks, Michael Dine, Jacques Distler, Michael Green, Brian Greene, Gary Horowitz, Clifford Johnson, Renata Kallosh, Juan Maldacena, Lubos Motl, Hermann Nicolai, Amanda Peet, Michael Peskin, Joe Polchinski, Lisa Randall, Martin Rees, John Schwarz, Steve Shenker, Paul Steinhardt, Kellogg Stelle, Andrew Strominger, Leonard Susskind, Cumrun Vafa, and Edward Witten for their time and patience. If we still disagree,

I hope it is clear that this book is not a final statement but a carefully structured argument, intended as a contribution to an ongoing conversation that has been undertaken with respect and out of admiration for their efforts. If the world turns out to be eleven-dimensional and supersymmetric, I will be the first to applaud their triumph. But for the present, I thank them in advance for allowing me to explain why, after a great deal of thought, I no longer believe this is likely.

This is not a scholarly history, but I do tell stories, and several friends and colleagues gave generously of their time to help me tell true stories rather than perpetuating myths. Julian Barbour, Joy Christian, Harry Collins, John Stachel, and Andrei Starinets gave me detailed scholarly notes on the whole manuscript. The mistakes that remain are, of course, my responsibility alone, as are the consequences of choices made to make the book as accessible as possible. Corrections and further thoughts will be posted on a Web page connected to this book. Other friends and family who read the manuscript and offered very helpful criticisms included Cliff Burgess, Howard Burton, Margaret Geller, Jaume Gomis, Dina Graser, Stuart Kauffman, Jaron Lanier, Janna Levin, João Magueijo, Patricia Marino, Fotini Markopoulou, Carlo Rovelli, Michael Smolin, Pauline Smolin, Roberto Mangabeira Unger, Antony Valentini, and Eric Weinstein. Chris Hull, Joe Polchinski, Pierre Ramond, Jorge Russo, Moshe Rozali, John Schwarz, Andrew Strominger, and Arkady Tseytlin also helped to clarify specific facts and issues.

For many years my research was comfortably supported by the National Science Foundation, for which I remain very thankful. But I was extraordinarily fortunate to encounter someone who asked me, "What would you really like to do? What is your most ambitious and crazy idea?" Then, unexpectedly and generously, Jeffrey Epstein gave me the chance to try to make good on my answers, and for this I will always be deeply grateful.

This book is partly about the values that should govern a scientific community, and I was lucky to learn mine from some of those who pioneered the search for quantum spacetime: Stanley Deser, David Finklestein, James Hartle, Chris Isham, and Roger Penrose. I would not have gotten anywhere on that search were it not for the collaboration and support of Abhay Ashtekar, Julian Barbour, Louis Crane, Ted Jacobson, and Carlo Rovelli. I am also indebted to my recent collaborators Stephon Alexander, Mohammad Ansari, Olaf Dreyer, Jerzy Kowalski-Glikman, João Magueijo, and especially Fotini Markopoulou, for continual criticisms and challenges that keep me honest and block any temptation to take myself too seriously. It must also be said that our work would not make sense without the wider community of physicists, mathematicians, and philosophers who ignore academic fashion to devote themselves to work on the foundational problems in physics. This book is dedicated to them, above all.

My work and life would be impoverished without the support of friends who enabled me to both do science and come to understand its larger context. These include Saint Clair Cemin, Jaron Lanier, Donna Moylan, Elizabeth Turk, and Melanie Walker.

Each book is written with the spirit of a place. For my first two, these were New York and London. This book carries the spirit of Toronto; Pico Iyer calls it the city of the future, and I count myself lucky to know why. For welcoming an immigrant at the uncertain moment of September 2001, I have to thank above all Dina Graser but also Charlie Tracy Macdougal, Olivia Mizzi, Hanna Sanchez, and the guys at the Outer Harbour Centreboard Club (if you didn't see me out on the water much last spring, this is why!).

For inviting me here, I have to thank Howard Burton and Mike Lazaridis. I know of no greater act of vision and support for science than their founding of the Perimeter Institute for Theoretical Physics. For their faith in the future of science and their continuing devotion to the institute's success, they deserve the highest praise that anyone who cares about science can give. I owe them enormous thanks for the opportunity they opened to me, both personally and scientifically.

For all the shared adventures and challenges of building that institute and community, all possible thanks to Clifford Burgess, Freddy Cachazo, Laurent Freidel, Jaume Gomis, Daniel Gottesman, Lucien Hardy, Justin Khoury, Raymond Laflamme, Fotini Markopoulou, Michele Mosca, Rob Myers, Thomas Thiemann, Antony Valentini, and others too numerous to name who risked their careers to contribute to this venture. And although it doesn't need to be said, let me emphasize that every word of this book is my own view and reflects in no way any official or unofficial views of Perimeter Institute, its scientists, or its founders. On the contrary, this book has been made possible by my membership in a community of scientists who celebrate honest scientific disagreement and who know that lively discussion need not get in the way of friendship or mutual support in our efforts to do science. Were there many more places like Perimeter, I wouldn't have felt the need to write this book.

Finally, to my parents, for their ongoing unconditional love and support, and to Dina, for everything that makes life a joy, which puts all that this book is about into its proper perspective.

Index